Aditya Aryasomayajula
Studying the electrotactic behavior of saos-2 cells
in non-homogenous electric field strength

TUD*press*

DRESDNER BEITRÄGE ZUR SENSORIK

Herausgegeben von Gerald Gerlach

Band 49

Aditya Aryasomayajula

Studying the electrotactic behavior of saos-2 cells in non-homogenous electric field strength

TUD*press*

2013

Die vorliegende Arbeit wurde am 1. Februar 2013 als Dissertation an der Fakultät Elektrotechnik und Informationstechnik der Technischen Universität Dresden eingereicht und am 14. Juni 2013 erfolgreich verteidigt.

Vorsitzender der Prüfungskommission:	Prof. Dr.-Ing. Klaus-Jürgen Wolter
Gutachter:	Prof. Dr.-Ing. habil. Gerald Gerlach
	Prof. Dr. med. habil. Richard Funk

Bibliografische Information der Deutschen Nationalbibliothek
Die Deutsche Nationalbibliothek verzeichnet diese Publikation in der Deutschen Nationalbibliografie; detaillierte bibliografische Daten sind im Internet über http://dnb.d-nb.de abrufbar.

Bibliographic information published by the Deutsche Nationalbibliothek
The Deutsche Nationalbibliothek lists this publication in the Deutsche Nationalbibliografie; detailed bibliographic data are available in the Internet at http://dnb.d-nb.de.

ISBN 978-3-944331-21-8

Verlag der Wissenschaften GmbH
Bergstr. 70 | D-01069 Dresden
Tel.: 0351/47 96 97 20 | Fax: 0351/47 96 08 19
http://www.tudpress.de

Gesetzt vom Autor.
Printed in Germany.

Editor's Preface

Life is thermodynamics. Everybody who has ever visited a doctor knows that he will be asked for his body's temperature, his blood pressure and, possibly, for reducing his weight (volume).
However, live is also electricity. Already in 1780, Luigi Galvani discovered that frog leg muscles contracted when he dissected them with a metal scalpel that was close to a static electricity source. This is caused by action potentials which occur not only in muscle cells but also in neurons, endocrine cells etc. Action potentials enable inter-cellular communication and activate intracellular processes. In 1928 E.J. Lund discovered that changes in voltage gradients often correlate with morphogenetic events during growth and patterning of animals. Several decades later it was discovered that the membranes of all cells contain ion channels based on pore-forming membrane proteins. Ion channels establish a resting membrane potential, control the flow of ions across cells, and cause the mentioned action potentials and other electrical signals by gating the flow of ions across the cell membrane. In 1952 Alan Hodgkin and Andrew Huxley were awarded the Nobel Prize for this discovery.
In the 1990s it was found that the application of physiological direct-current electrical fields (DC-EF) lead to a movement of cells. This phenomenon is called electrotaxis or galvanotaxis. Several cell types migrate to the cathode (e.g. fibroblasts and many epithelial cell types), whereas other cell types move to the anode (e.g. human vascular endothelial cells). It was shown that both speed and movement directions are voltage-dependent. Recent studies suppose that DC-EFs influence wound healing, stem cell differentiation and many other biological processes.
So far most experiments regarding electrotaxis of cells used AC fields whereas just a few numbers of studies dealt with DC-EFs. The author of this book was the very first who extended the approach of applying electrical voltage to cells by studying cell behaviour under the influence of electrical field gradients. His work is based on the development of a cell chamber with an electrode geometry enabling the supply of EF gradients on cells including the fast switching of field gradients. First experiments have shown that his measuring set-up is able to study the influence of EF gradients on cell behaviour, e.g. on cell motility. His findings will enable further activities to explain EF effects on cells in particular and in cell biology in general.

Since this entire field seems to be a very attractive one in the future development of cell biology, it makes me sure that this volume of the book series "Dresden Contributions to Sensor Technology" will earn that amount of kind attention it deserves.

Dresden, June 2013 Gerald Gerlach

Preface of Author

This work was carried out as a member of the DFG Graduate College "Nano and biotechnologies for the packaging of electronic systems" at the institute of solid state physics and anatomy at the Technical University of Dresden. The device described in this work was fabricated at the solid state physics laboratory and the biological studies on cells were carried out at the institute for anatomy. Previously the cells were studied in homogenous electric field strength. A novel device was fabricated and tested for studying cells in non-homogenous electric field strength.

The development of this device required inter-disciplinary research between engineering and biology. This was possible only with the help of colleagues and friends in both the fields whom I hereby thank:

- Prof. G. Gerlach for scientific expertise and support throughout my work. Also for giving me an opportunity to work on this project and for the moral support.
- Prof. R. H. W. Funk for all the useful discussions and support in cell biology.
- Dr. S. Perike and Dr. J. Derix for all the helpful discussions and comments on my work.
- D. Koslowski, T. Hoche and K. Perike for carrying out the cell culture.
- My colleagues: A. Schroeter, R. Hensel, M. Luniak, V. Schultz and J. Posseckardt
- All the colleagues at the institute for anatomy for their help in microscopy and cell culture.
- S. Kostka, U. Lehmann, S. Herbts and all the other colleagues at the institute for solid state physics department for their help and support.
- All the colleagues of the DFG Graduate College for their support.
- Prof. J. Wolter for allowing me to use equipment in their institute.
- My sister L. Arysomayajula and my family for their continued support and motivation.

Dresden, January 2013. Aditya Aryasomayajula

Abstract

Cell motility is a fundamental process found in all kind of living cells. Ion transporter proteins located on the cell membrane produce endogenous electric fields (EFs) which are crucial for many biological processes like nerve regeneration and embryonic development. Recently, these EFs were found to play a crucial role in the wound healing process. An electric field is generated near the wound location which the neighboring cells use as a cue to migrate towards the wound and start the healing process. Such a phenomenon of directed cell migration under the influence of electric field is called electrotaxis or galvanotaxis. A standard electrotaxis experiment consists of an electrotaxis chamber, a dc power source, and agar bridges. The cells are cultured in this electrotaxis chamber and the migration behavior is studied under a time-lapse microscope. The electric field strength (EFS) used for such an experiment are uniform and homogeneous. The EFS around a wound location is generally multi-field and non-homogenous.

In order to study more closely the wound healing process in vitro, a device which can generate non-homogenous EFS should be used. In this work, we study the cells in a device which is capable of producing both homogenous and non-homogenous EFS. The device is simple to fabricate and live imaging of cells can be performed. Polydimethylsiloxane (PDMS) is used to fabricate the micro-channels which is sealed by a 4-well cell culture well. The cross shape of the micro-channel allows the use of multiple EFS which produces non-homogenous EFS. The EFS inside the device is simulated and verified experimentally.

We observed that the cells in non-homogenous EFS migrated in the resultant direction of the EFS. This finding is particularly important to control the migration direction of cells by manipulating the EFS inside the electrotaxis chamber.

Table of contents

1 INTRODUCTION

1.1 Motivation

Endogenous electric fields (*EFs*) are produced naturally in living cells (Figure 1). In 1780, Luigi Galvani when dissecting a freshly killed frog accidentally touched the nerve in the frog's leg with a metal scalpel that was close to a static electricity source. He noticed that the frog's leg twitched. This observation led him to conclude the existence of "animal electricity" [1]. A few years later, he made another accidental discovery. He hung some frog's legs by copper hooks to an iron railing and when the frog's legs touched the iron railing, the muscles twitched although there was no source of static electricity nearby. Galvani concluded that there was a potential difference between the frog's legs and the iron railing. Later in 1850, DuBois-Reymond measured bio-electric currents of 1 μA from a wound proving the existence of these *EFs* [2]. His findings were the foundation for modern electrophysiology, the study of electrical properties of cells and tissues. Since then, a lot of progress has been made in understanding more about such behavior of cells. A big break-through was the measurement of ionic currents flowing out of ion channels from cells by Erwin Neher and Bert Sakmann in the early 1980s [3]. Very recently (2005), Zhao et al. [4] showed that phosphatidylinositol-3-OH kinase-γ and phosphatase and tensin homolog are the two genes involved in directional migration of cells. Many other groups have studied the electrical properties of cells (see chapter 3). But in all these studies, the cells were exposed to uniform electric field strengths (*EFS*). Also, the electric field characterization inside the devices used for such studies were poorly understood. In vivo conditions suggest that these cells are exposed to non-homogenous *EFS*. This work studies the effect of non-homogenous *EFS* on cells and the dependence of device channel parameters to *EFS*s.

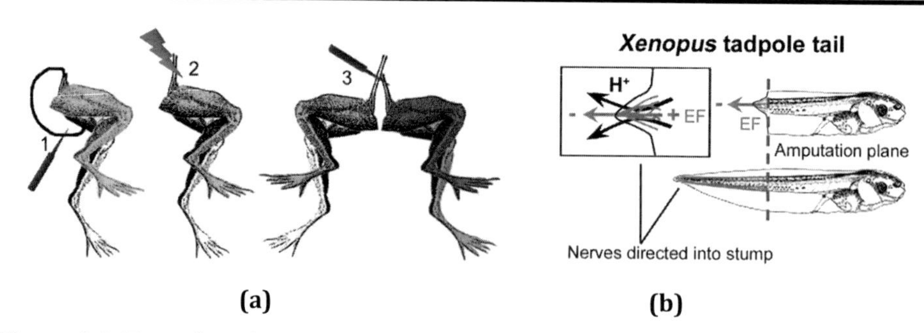

Figure 1.1: Examples of endogenous electric fields in animals. (a) Twitching of frog's legs in Luigi Galvani's classic experiment [1]. (b) Regeneration of Xenopus tadpole tail and spinal cord by *EFs* [5]

EFs play a crucial role in many biological functions. These *EFs* are involved in regeneration of nerve cells and in embryo formation [5]. Recent observations have shown that these *EFs* play an important role in wound healing processes [4]. *EFS* of 0.02 to 0.10 V/mm are produced at the wound location [6, 7]. During the second stage of wound healing process, called hemostasis, the cells use these *EFs* as directional cues to migrate towards the wound location. This phenomenon of directed migration of living cells under the influence of *EFS* is known as *electrotaxis* or *galvanotaxis*. Studying the electrotaxis behavior of cells helps us to better understand the wound healing process and other important biological functions.

Another area where these *EFS* have gained considerable interest is in the stem cell differentiation [8]. Stem cells are cells which can differentiate into different specialized cell types and can divide to renew themselves. For example a stem cell under certain conditions can be differentiated into a liver cell. These liver cells can then be used to repair damaged liver cells in the body. Under what conditions these stem cells differentiate into different cells is still not fully understood. Recent findings have shown the use of *EFs* to differentiate the stem cells into specialized cardiac cells using electric fields [8]. By studying the effect of *EFS* on cells, stem differentiation could be better understood.

1.2 Novelty of work

Majority of the electrotaxis experiments studied use direct current electric field (*dcEF*) compared to alternating current (ac) since the *in-vivo EFs* are believed to have a nature of *dcEF* [5]. When using *dcEF* the electrodes generate toxic products which are harmful for the cells. In order to overcome this difficulty, agar salt bridges are used to transfer the current from the electrodes to the cell chamber. These agar salt bridges slow down the movement of these toxic products. Electrotactic phenomenon has been studied for the past 20 years. Most of these studies are conducted in a homogenous *dcEF* [see chapter 3]. In a homogeneous *dcEF* experiment, the cells experience uniform *EFS*. However there exists a non-homogenous *dcEF* near a wound location. In order to study the electrotaxis phenomenon during a wound healing process in vitro, it is necessary to use non-homogeneous *dcEF*. To our knowledge, this is the first time the electrotaxis phenomenon of cells is studied in non-homogenous *EFS*. A novel design for an electrotaxis device for studying cells in both homogenous and non-homogenous *dcEFs* is presented here. *EFS* in both homogeneous and non-homogeneous *EFS* are simulated using finite element simulation software and then verified experimentally. The device is easy to fabricate and live cell imaging can be performed directly under a microscope.

1.3 Hypothesis

The main hypothesis of this work is to understand the migration behavior of cells in non-homogenous *EFS*. The migration direction of cells in non-homogenous *EFS* would depend on the resultant direction of the *EFS*. This result would be useful in controlling the migration direction of cells just by applying the *EFS* in the required direction.

1.4 Outline of the thesis

In the following chapters of this thesis the construction of device to study electrotaxis and characterization of *EFS* inside this device are explained. In addition, the results of migration behavior of cells in non-homogeneous *EFS* are also discussed.

To better understand the working principle of the electrotaxis device, basic concepts in electric engineering about *EFS* are explained in chapter 2. It also consists of the description of a standard electrotaxis experiment. Basic terminology and science of living cells are described which include cell anatomy, cell signaling process and electrotaxis phenomenon. These concepts are required for better understanding of the principles and procedures described in the later chapters.

Chapter 3 reviews the state of the art in the field of electrotaxis phenomenon. This chapter highlights the work done so far in this field. The studies are listed in a chronological order starting from the 1850s to present time to understand how the field of electrotaxis started and how much far have we come to understand this phenomenon. Important discoveries are highlighted to see the transition from one age to the next.

The steps involved in fabrication of the device used for studying the electrotaxis phenomenon as well as the cell culture procedures are explained in detail in chapter 4. The materials used and the working principles of different microscopes are described. The quantification of cell migration data is also explained in this chapter.

The characterization of *EFS* inside the device is described in chapter 5. First, homogenous *EFS* dependence is analyzed on different channel dimensions. The *EFS* distribution for non-homogenous *EFS* is then studied for two different configurations. The *EFS* were first simulated using finite element software and these results were verified experimentally for both homogenous and non-homogenous *EFS*. The simulation and experimental results matched closely. So the simulated *EFS* values were used to quantify the migration direction of the cells in non-homogenous *EFS*.

In chapter 6, the results of cell migration behavior on both homogenous and non-homogenous *EFS* are presented. First the biocompatibility of the device is tested by culturing the cells for 24 hours and analyzing the cell attachment, spreading and growth inside the device. A control experiment is then carried out to show the migration behavior of cells without *EFS* is random. The cells were then exposed to homogenous *EFS* and they showed directional migration towards the anode. In non-homogenous *EFS*, the cells migrated in the resultant direction of the *EFS*.

The conclusions and future direction of this work are discussed in chapter 7. The main conclusion of proving the hypothesis and the possible use of this device to study the migration behavior of cells in non-homogenous *EFS* is explained. The possible use of this device in the future could be to study stem cell differentiation.

2 Research Background

This chapter explains the theoretical basics in electrical engineering and cell and molecular biology. The section 2.1 of this chapter explains the basic principles of electrical engineering used in this work. It comprises the definition of *EFS* and its properties as well as calculating resistance in an electrical system. Section 2.2 describes the standard technique used for carrying out conventional electrotaxis experiments. The third section contains the basic concepts and mechanisms involved in cell and molecular biology. The cell anatomy and the process of signaling pathways as well as mechanisms involved in electrotaxis phenomena are explained.

2.1 Electric field strength

EFS is defined as the intensity of an electric field by the force F experienced by a stationary positive unit point charge, q, at a particular location. It can also be described as the negative gradient of the electrical potential, Φ, at this location:

$$EFS = -\nabla\Phi, \tag{2. 1}$$

where $\nabla = (d/d\,\vec{r})$ is the gradient with respect to the space vector $\vec{r}$. If the electric field is homogenous then *EFS* simplifies to

$$EFS = \frac{V}{l}, \tag{2. 2}$$

where V is the voltage applied to the system and l the distance between the measured points. *EFS* has a unit of V/mm.

In biological language, *EFS* is also called physiological strength. It is defined as the voltage required for stimulating a living cell electrically. Typical values of this physiological strength vary between 0.02 and 0.10 V/mm depending on the type of cell [6, 7]. This physiological strength is one of the main factors in designing a micro-fluidic device for studying these cells.

One of the fundamental laws in the field of electrical engineering is the Ohm's law. It gives a relation between voltage, current and resistance, R, if voltage V and current I are proportional and in phase:

$$I = \frac{V}{R}. \tag{2. 3}$$

Furthermore, the resistance, R, of an element is defined as the opposition to the passage of current through it. The resistance depends on the material and geometry of the element. It can be expressed as

$$R = \frac{\rho . l}{A}, \tag{2. 4}$$

where ρ is the resistivity of the material [Ωm], l is the length [m] and A is the cross sectional area [m^2]. The resistance in a system plays an important role in determining the amount of current that can pass through the system. Because the resistance is directly proportional to the voltage in the system, for higher resistance a higher voltage is required to get the same current.

Figure 2.1 shows the relation when resistors are connected in series and parallel in a circuit.

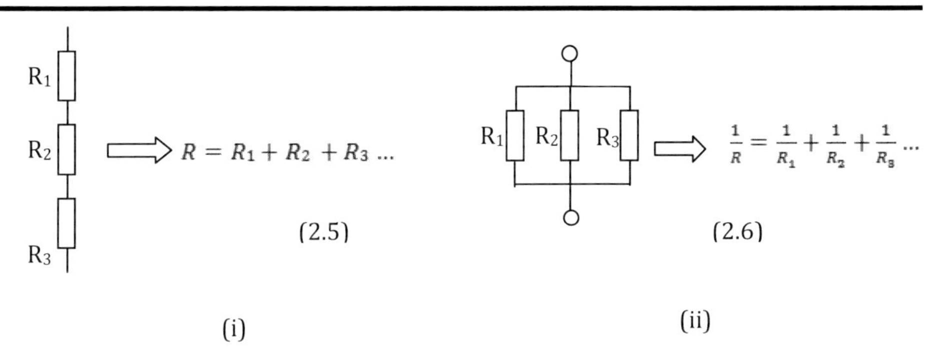

Figure 2.1: Resistors in a circuit: (i) resistors in series, (ii) resistors in parallel

2.2 Standard electrotaxis experiment

Electrotaxis is a phenomenon in which certain cells migrate under an influence of a *dcEF* towards an electrode. Depending on the type of the cell, the cell would either move towards the cathode or anode. Different types of cells have been found to show this behavior [6]. This migration behavior is important for instance during the wound healing process [9]. Figure 2.2a shows the standard experimental setup used for electrotaxis experiment [5]. It consists of an electrotaxis chamber in which the cells are cultured over a period of 24 h. In order to study the migration behavior of cells, a *dcEF* is applied using a dc power source. The problem with *dcEFs* is that toxic products are generated at the electrodes which are harmful for the cells. In order to avoid these toxic products from entering the electrotaxis chamber, the current is transferred from the dc power supply first to a saline-filled beaker (typically 4 % NaCl solution) by an Ag/AgCl electrode. This current is then conducted into the electrotaxis chamber by means of an agar bridge setup as shown in the fig. 2.2. The agar bridges slow down the diffusion of toxic products into the electrotaxis chamber. On the other hand these agar bridges contribute a considerable resistance to the system. This whole setup is placed inside an incubator which is maintained at 37 °C and 5 % CO_2. This incubator is housed inside a microscope for live cell imaging. A typical electrotaxis experiment is run for 5...6 h depending on the type of the cell being studied. Song et. al recently published the standard protocol for carrying out such an experiment [10].

A simplified electrical circuit diagram of such an experimental setup is shown in figure 2.2b. In this figure, R_0, R_1 and R_2 are the resistance of the platinum wire, agar bridges and cell culture medium, respectively. The total resistance of the system can be calculated from the equation 2.5 as the sum of all the resistances:

$$R = 2\,R_1 + R_2 \tag{2.6}$$

Where R is the total resistance and $R_0 << R_{1,2}$

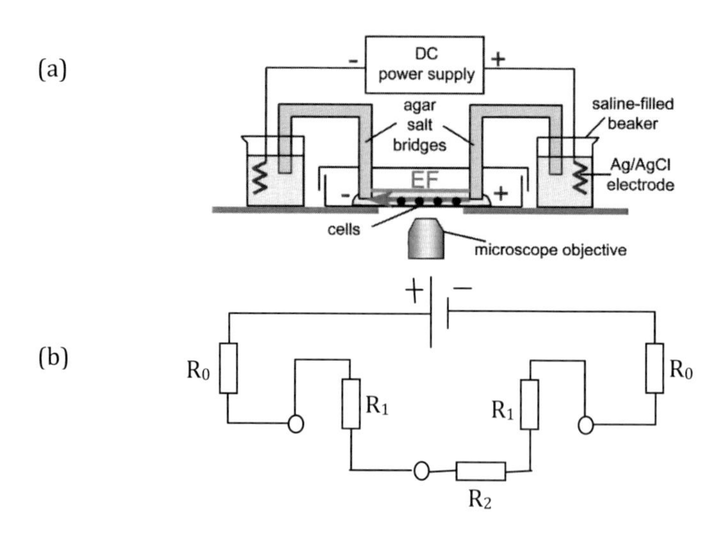

Figure 2.2: Schematics of a standard electrotaxis experiment: (a) standard electrotaxis experimental setup [5], (b) corresponding electrical circuit diagram

The electrotaxis chamber consists of a straight channel, usually 0.4~0.6 mm in height. This channel can either be engraved on a glass petri dish or fabricated my soft lithography technique by using the polymer polydimethylsiloxane (PDMS).

Another most common electrotaxis chamber used in research can be obtained commercially from the firm IBIDI Gmbh (Figure 2.3) [11]. This slide consists of a micro-channel 50 x 5 x 0.4 mm in length, width and height, respectively. The two ends of the channel are connected to reservoirs which are filled with the cell culture medium. The cells are seeded inside this channel and *dcEF* is applied.

The *EFS* inside such an electrotaxis chamber is always homogenous. The cells experience an uniform *EFS* when migrating. This is because the channel dimensions are constant and *dcEF* is applied only in one direction.

The *EFS* or physiological strength applied in the channel depends on the type of cell under study. Different cells have different *EFS* at which they are stimulated [6]. After the application of *EFS*, the cells are observed under a time-lapse microscope for a period of time. The migration data is then quantified using tracking software where the nucleus of the cell is tracked for the period of observation. This simple method provides a powerful insight into the migration behavior of cells under *EFS*.

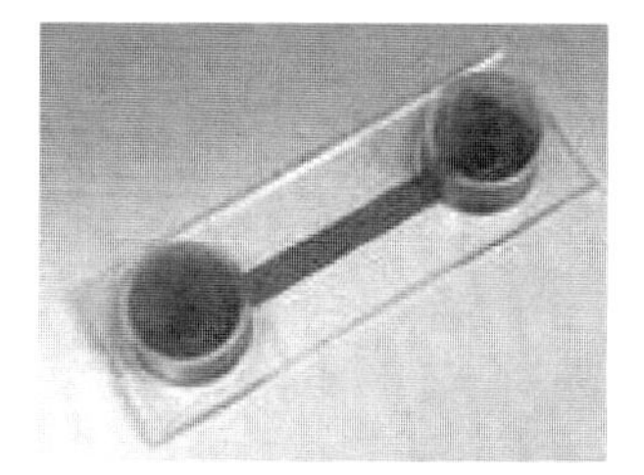

Figure 2.3: IBIDI micro-slide for electrotaxis experiment [11]

2.3 Cell and molecular biology

2.3.1 Cell anatomy

Figure 2.4 shows the anatomy of a general animal cell. The main components of a cell are [12]:

- Nucleus: stores the genetic material in an animal cell.
- Endoplasmic reticulum: network of membranous sacs and tubes. It is active in membrane synthesis and metabolic processes and consists of both rough and smooth endoplasmic reticulums.
- Flagellum: motility structure present in some animal cells.
- Centrosome: region where the cell's microtubuli are initiated.
- Mitochondrion: organelle where most of the energy of the cell is produced.
- Golgi apparatus: organelle active in synthesis, sorting, modification and secretion of cell products.
- Plasma membrane: membrane enclosing the cell.

The structural elements in a cell include microtubuli, microfilament and intermediate filaments as shown in Figure 2.5. These structures help the cell to maintain its structural integrity and are also required for cell migration.

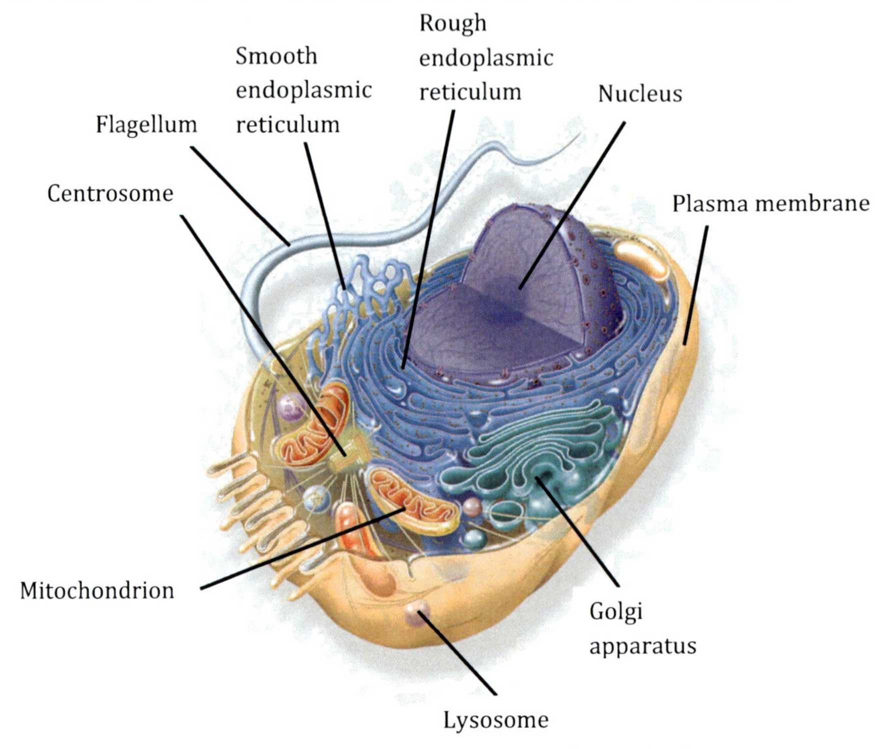

Figure 2.4: Sectioned view of a generalized cell [12]

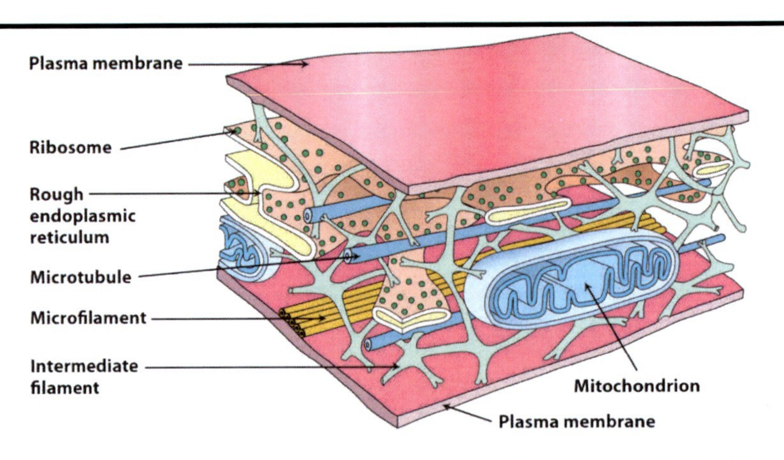

Figure 2.5: Schematic diagram of the structural elements in cells [12]

2.3.2 Cell migration mechanism

Cell motility is a fundamental process found in all kinds of cells. Migration of cells is important for many pathological and physiological processes [9]. In animal cells, cell migration starts from embryonic development [5] and are involved in immunological responses [13]. Cell migration plays a vital role in physiological processes like wound healing [4], regeneration of tissues [14] and tumor metastasis [15].

The failure of cells to migrate or migration of cells to a wrong location can result in abnormalities and have life-threatening implications. For example, abnormal migration of neural cells could lead to congenital defects in the brain development leading to mental disorders [16].

Most of the cells follow a common migration mechanism. The push-pull model is widely used to explain the cell migration phenomenon [17]. This model explains the cell migration by three sequential processes:

a. *Polarization of the leading edge*

Cell migration begins with the cell's response to an external signal that leads to the polarization of the cell. There are different ways how a cell can respond to external signals that initiate and promote cell migration. These external signals could be a chemical gradient, substrate morphology or electrical signals in its environment. A polarized cell is a spatiotemporal reorganization of actin filaments which is morphologically and functionally different from a normal cell [16]. Polarization of such a migrating cell is achieved by the protrusions of the membrane in the direction of the migration at the leading edge and dislodgement of cell substrate at the rear edge [17]. At the leading edge, the membrane protrusions are driven by the continuous reorganization and turnover of the actin cytoskeleton. Actin filament polymerization drives the extension of sheet and rod-like protrusions at the leading edge of the cell [18]. At the rear edge, actin interacts with myosin to contract the trailing cell portion [19]. Figure 2.6 shows the reorganization of actin filaments at the leading and trailing end of a migrating cell.

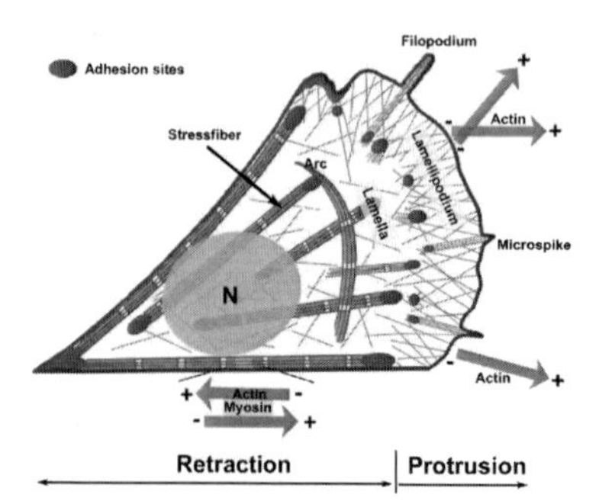

Figure 2.6: Reorganization of actin filaments at the leading and trailing edges of a cell [17]

b. *Adhesion turnover and focal contacts*

Effective migration is achieved by adhesion of membrane protrusions to the substrate. Receptor proteins such as integrins are used by the membrane protrusions to adhere to the substrate which closely interact with the actin cytoskeleton (Figure 2.7). The cell uses focal adhesion contacts for molecular interactions with the extra cellular matrix. These focal adhesion contacts are formed by the activity of focal adhesion kinase (FAK) which is found at the leading edge of the cell [18].

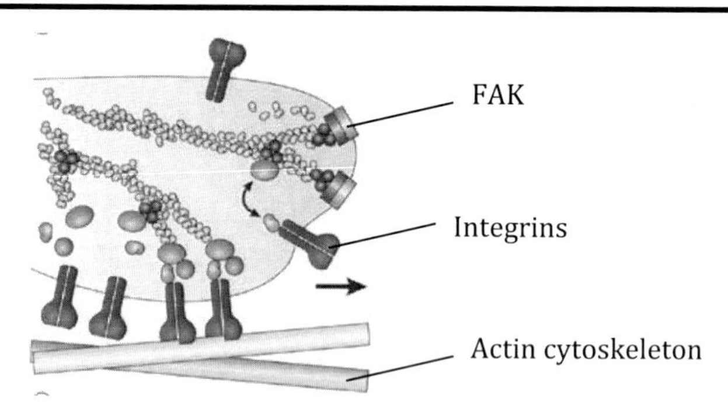

Figure 2.7: Cell protrusions at the leading edge [18]

c. *Detachment of the trailing edge*

At the trailing edge, detachment of focal adhesion contacts occurs through several mechanisms which include the contraction of the cortical actomyosin network at the cell pole [19] and calcium-dependent proteolytic enzyme calpain activity [20]. In cell migration, calpain activity in necessary to discharge the integrin contacts at the rear trail [19]. This phenomenon is due to the cleavage action of calpain on FAK.

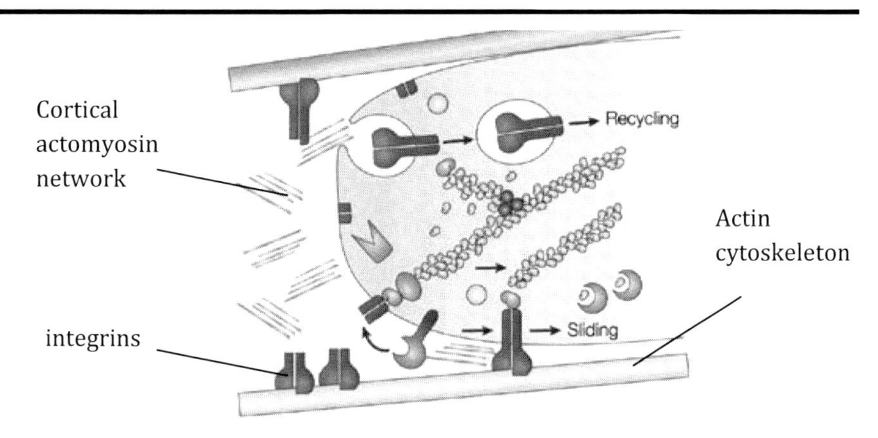

Figure 2.8: Detachment of the trailing edge [19]

2.3.3 Ion channels

Ion channels regulate the flow of ions into and out of the cell membrane. They are integral membrane proteins present in the plasma membrane of the cells. Ion channels are very important for maintaining the osmotic pressure inside the cell and for cell signaling. Ion channels may generally be classified based on the type of gating and the species of ions passing through the gates:

- Voltage-gated channels: they allow the flow of only one type of ion through it. They may be activated by a voltage signal. Known voltage-gated ion channels are sodium, calcium and potassium.
- Ligand-gated channels: They are activated by specific ligand molecules when they bind to the exterior domain of the receptor protein. A conformational change in the structure of the protein opens the gate for the ion to pass through such channels. They are mostly used for signal transmission in cells.

The voltage-gated calcium ion channels which regulate the flow of calcium ions in a cell are particularly important with respect to the motility of the cell which is a Ca^{2+}-dependent process. They mediate calcium flow into the cell in response to a signal and control intracellular processes like contraction, secretion, neurotransmission and gene expression. Their activity is essential to transmit an external electrical signal to physiological events inside the cell [21]. Ca^{2+} modulates the turnover of actin filaments. There is less concentration of Ca^{2+} ions at the leading edge which localizes the actin polymerization [20].

2.3.4 Cell signaling pathway

Cell signaling is important for many biological and physiological processes occurring inside a cell. It can broadly be defined as the transmission of signals from the outside of the cell membrane to the inside of the cell through trans-membrane proteins. The term *signal transduction cascade* is used for how a cell senses an external signal stimulus and processes that signal by additional internal layers of signals. A simplified model of the signal transduction pathway in cells is illustrated in Figure 2.9. It consists of the following steps [12]:

- *Signal*: The first step is the generation of a signal. The stimulus could be any biological or physiological process like a wound. This stimulus signal is called a primary messenger.
- *Reception*: The second step is the reception of the primary signal by membrane proteins. These membrane proteins are responsible for transferring information from the external environment to the cellular interior. Conformational changes in the structure of the receptor proteins are responsible for signal transmission. When a signal molecule (also called a ligand) binds to the extracellular domain portion of the receptor protein, a conformational change occurs. These conformational changes relay all the way to the intracellular membrane.
- *Transduction*: The third step is the relay of signal by second messengers. The information is converted into small molecules, called second messengers, whose concentration changes as a result of stimulus. These molecules are responsible for amplification of the incoming signal. Examples of such second messengers include cAMP, cGMP, Ca^{2+} ions.
- *Response*: Upon transduction to secondary messengers, multiple subsequent signaling steps can take place to generate a number of cell responses in a highly coordinated and regulated fashion.
- *Termination*: Through feedback regulation, the responses generated by the cell can stop the upstream regulation like transduction, reception and generation of signal.

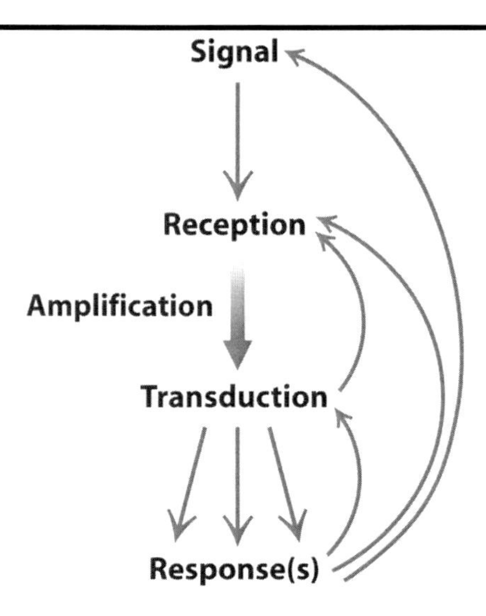

Figure 2.9: Signaling transduction model in living cells [12]

Every biological activity is a signal transduction pathway or can be modified by one. Figure 2.10a shows a general signaling pathway found in most of the cellular systems. A seven trans-membrane (7TM) receptor, seen as embedded into the cell membrane, binds to a ligand (5-HT in this case). A conformational change is transmitted to the cytoplasmic side of the membrane. Another protein is associated with the cytoplasmic side of the cell called the G-protein. The conformational change in the 7TM receptor activates this G-protein consisting of three subunits called α, β and γ. The G-protein activation allows the exchange of the GDP nucleotide to GTP. As this occurs, the β and γ subunits dissociate from the G-protein leaving a G-α complex. This G-α complex transmits the signal that a ligand is bound to the protein. For a single binding of ligand hundreds of these G-α complexes are generated and this is how signal amplification is carried out. Once GTP is generated it binds to another protein called adenylate cyclase. When GTP binds to adenylate cyclase it activates it which is used to generate phospholipase C (PLC). PLC further activates phosphatidylinositol 4,5-bisphosphate (PIP2). Phosphorylated PIP2 generates downstream molecules of inositoltrisphosphat (IP3) and diacylglycerine (DAG).

Figure 2.10b shows how IP3 can further activate protein kinase C (PKC) which is a common enzyme for most of the process used by cells. PKC activation requires the binding of two DAG molecules and two Ca^{2+} ions. The Ca^{2+} ions are generated by activating the IP3 receptor through IP3 molecules. This specific pathway thus shows the importance of Ca^{2+} ions in the signaling step.

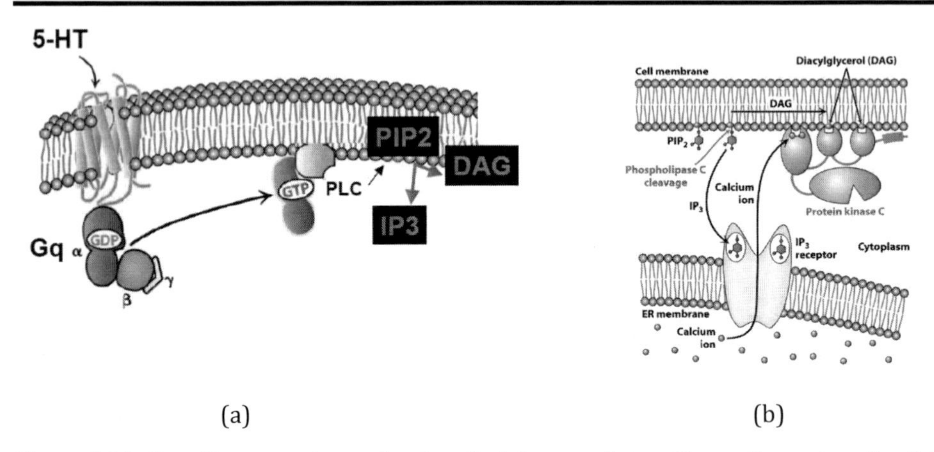

(a) (b)

Figure 2.10: Signaling cascade mechanism []: (a) general signaling pathway in cells, (b) activation of PKC by IP3 molecule [12]

2.3.5 Electrotaxis

Electrotaxis is defined as the directed migration of living cells under the influence of a *dcEF*. Depending on the type of cell, they can migrate either towards the cathode or the anode [6]. Electrotaxis should undergo the same cytoskeleton rearrangement as observed for normal cell migration, although the mechanisms controlling the directionality and speed may differ. There are currently four factors that may cause the electrotaxis effect in cells.

a) *Ca^{2+} ion distribution inside the cell*

In most of the studies conducted on electrotaxis behavior of cells, changes in concentration of calcium ion ([Ca^{2+}]) mediated electrotaxis. This agrees to the fact that the initial response to *dcEFs* is fast and would involve small molecules. An increase in [Ca^{2+}] was observed when *dcEF* was applied to mouse embryo fibroblasts [22, 23]. When a *dcEF* is applied, the rise in [Ca^{2+}] in a given part of the cell causes contraction on that part of the cell and the opposite side is protruded. Generally, in cathode-moving cells, a higher [Ca^{2+}] is observed at the rear end [24]. This would start a "push-pull" movement [21] of the cell which could involve two mechanisms:

- Actin polymerization/depolymerization. The main driving force for cell migration is the elongation of the actin filament at the leading edge. Gelsolin induces the generation of free barbed ends of actin filaments. At the protruding end of the cell, the reduction of [Ca^{2+}] releases gelsolin from these barbed ends which mediates polymerization of actin filaments in that direction. At the rear end of the cell, increased [Ca^{2+}] may induce depolymerization of actin filaments facilitating in detachment of the cell.

- Actinmyocin contractility. Myosin II is a common motor protein found in muscle and non-muscle cells and is regulated by [Ca^{2+}]. The phosphorylation of myosin light-chain kinase stimulates the actin-activated myosin which plays a major role in cell contraction. Also, increase in [Ca^{2+}] triggers contraction of non-muscle cells.

In order to confirm the importance of [Ca^{2+}] on electrotaxis of cells, experiments were performed where calcium ion channel blockers like Co^{2+} or D600 were used and in most cases electrotaxis was observed to be inhibited [25].

b) *Voltage-gated Na^{+} channels*

The effect of voltage-gated Na^{+}-channels (VGNC) is still not fully understood [26]. One possible explanation is the indirect involvement of Ca^{2+} ions by VGNC. Na^{+}-influx through VGSCs could increase the concentration of [Ca^{2+}] locally by altering the release and uptake of Ca^{2+} intracellular stores through disruption of normal pH-regulating mechanisms [26].

c) *Surface charge of the cell*

There exists a surface charge in many cells due to the presence of charged residues in the plasma membrane. When a *dcEF* is applied, this charge cloud could move thereby creating a spatial variation in membrane potential. This spatial variation could in turn affect VGNC and other voltage-dependent ion channels.

d) *Growth factors and protein kinases*

Several other mechanisms have been found to influence electrotaxis. These include protein kinases such as protein kinase C [25] as well as growth factors and their receptors. These growth factors include epidermal growth factors, fibroblast growth factors and transforming growth factors [27]. However, much work needs to be done to fully understand the influence of these growth factors and protein kinases on electrotaxis.

3 State of the art: Electrotaxis

This chapter highlights the work done so far in the field of electrotaxis. The findings are listed in a chronological order starting from the 1850s to present time to see how the field of electrotaxis started and how far have we come to understand this phenomenon. The development of technology and knowledge from other fields like biochemistry greatly helped in solving the mysteries of the electrotaxis phenomenon. Important discoveries are highlighted to see the transition from one age to the next.

3.1 Introduction

Cell motility is a fundamental process found in all living cells. Directed cell migration is important for embryonic development and is required throughout the life. The failure of cell to migrate or the migration of cells to wrong locations can have life-threatening implications. There are different cues a cell can utilize for directed migration. These include:

- Chemotaxis-movement of cells due to chemical stimuli.
- Hapotaxis-response to graded adhesion in the underlying substrate or other guidance cues anchored in the extracellular matrix.
- Durotaxis-response to mechanical signals in the environment.
- Electrotaxis-based on migration of cells under the influence of an electric field (EF).

This review will focus on the devices used for studying the electrotaxis phenomenon and the results of the work done so far in this field. In the years from 1700 to 1950 the phenomenon was first reported and early discoveries were made. Sufficient knowledge in cell biology was not available during this period to explain the findings made. In the next three decades (1950-1980) technology and knowledge in cell biology helped to study more cells exhibiting this phenomenon and some important findings were reported. In the years between 1980 and 2005, the electrotaxis phenomenon was widely studied and one of the most important discoveries regarding the genes responsible for this phenomenon was found out. Researches all over the world started fabricating new devices using different technologies to study cells exhibiting this phenomenon in the next few years (ca. 2005-2010). The last years since 2010 saw the first applications of electrotaxis phenomenon in clinical therapies and stem cell research.

3.2 1700-1950

One of the first evidence for the influence of current on the human body dates back to mid 1700 when L'Abbe Jean-Antoine Nollet used the electricity from Layden jar to pass current through 180 of the king's men, which made them leap simultaneously [28]. Around the same time Galvani in his famous experiment showed the existence of bio-electric currents in living cells when the legs of a frog twitched when he touched the nerve muscle with a metal scalpel [1]. Later in 1850, DuBois-Reymond measured bio-electric currents of 1 μA from a wound proving the existence of these *EFs* [2]. His findings were the foundation for modern electrophysiology, i.e. the study of electrical properties of cells and tissues

One of the first published works on electrotaxis phenomenon on cells was by Verworn [29] in 1896. He observed that, when an amoeba is exposed to weak electric currents, it orients and moves towards the cathode. For stronger electric fields, the anodal side of the amoeba is contracted. He was not able to explain his observation.

In 1900 Carlgren [30] hypothesized that the anodal contraction is due to the negative charge on the solid portion which leads to electro endo-osmotic movement of fluid from the anodal to the cathodal side. Another explanation for this phenomenon was given by Hirschfield (1909) [31] and McClendon (1911) [32] where they argued that the current causes relative reduction in the surface tension on the cathodal side of the amoeba which causes the movement towards the cathode. In 1926 Luce proposed that the current causes the plasma gel on the cathodal side to become rigid which decreases the elastic strength in this region [33]. His explanation was based on the results obtained by Greeley [34] in 1904 where he experimentally showed that in an amoeba the current causes the anodal side of the protoplasm to coagulate and the cathodal side to liquefy. This conception was opposed by Mast [35] where he showed in his paper in 1931 that electric current always leads to a decrease in viscosity in some portions of the cell and simultaneous increase in others. He further explained that the increase or decrease in viscosity is due to localization of accumulated negative and positive ions, respectively. In 1948 Miller and Goldston [36] studied the electrotactic behavior of paramecia by using a balanced square waves they were able to stimulate paramecia and induced directional migration. Figure 3.1 shows the set-up used to produce the balanced square waves via platinum electrodes for stimulation. The paramecia were cultured on a glass slide and the whole device was enclosed inside a glass back plate. Definite current densities were produced using this apparatus at various frequencies. During this period most of the studies reported could not explain the phenomenon of electrotaxis because of the lack of knowledge and technology.

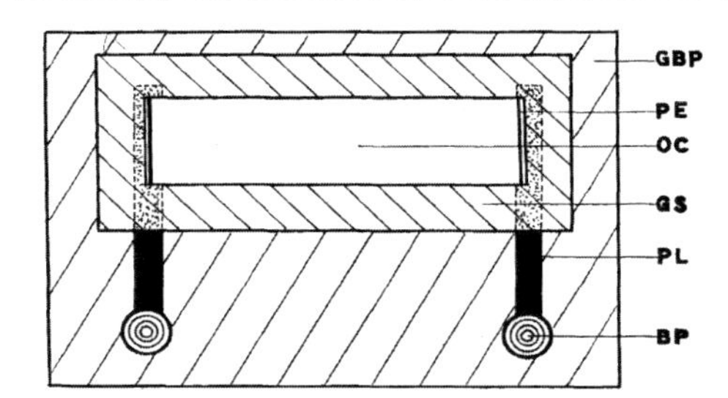

Figure 3.1: Apparatus for producing balanced square waves for stimulation of paramecia. GBP glass base plate; PE platinum electrode; OC observation cell; GS glass siding; PL platinum leads; BP binding post [36]

3.3 1950-1980

The response of organisms to electric field encouraged many other researchers to study more varied species and cell types for this phenomenon. In this period of time many new species and cell types were reported to exhibit electrotaxis behavior.

Negative electrotaxis was observed in several species of slime molds by Watanabe et al. [37]. Anderson [38] in 1951 published his work on electrotactic behavior of slime molds. He studied the migration of physarum polycephalum on different substrates like baked clay, paper and agar with current densities of 1 $\mu A/mm^2$. When the current densities were increased to more than 8 $\mu A/mm^2$ injuries were produced. He concluded that the electrotactic behavior is not due to electrode products. He also demonstrated that the electrotactic behavior is not due to gradients in substratum, pH changes in the mold and electroosmosis or endosmosis. He further observed that, by adding salts to the agar substrate, the change of cations had more effect on electrotactic behavior than anions. His work showed that the negative electrotactic response is due to inhibition of migration on the anodal side of the plasmodium. At that time the anodal migration was not fully understood. Fukushima et al [39] in 1953 showed that the granulocytes move towards the anode at high pH values and towards the cathode at low pH values.

The major problem in all these experiments was that the current density that passed through the organism observed was not known, only the current directed through the medium was measured. In 1974, Jaffe and Nuccitelli [40] measured steady extracellular currents from individual cells using the vibrating probe method. By this non-invasive technique they were able to detect current densities of 10 nA/m^2 in serum with a 30 μm diameter probe vibrating at 200 Hz (Figure 3.2). The probe directly measures the voltage difference between two extreme points of it vibration which is around 30 μm apart. The *EFS* is nearly constant for this distance, it is approximated as the voltage difference divided by this distance. The current density can then be calculated by multiplying this *EFS* with the conductivity of the medium.

The vibrating probe was used by Borgens et al. [41] to show steady currents leaving the stumps of regenerating newt limbs for 5 to 10 days after amputation. These currents were not affected by section of the main nerves of the limb. Jaffe and Nuccitelli [40] found that the stump of a salamander limb drives a direct current of amplitudes up to 100 $\mu A/cm^2$ outwards for about ten days. Borgens et al. [42] also found that these currents help in regeneration if they were introduced artificially into non-regenerating organisms and reversal of currents in salamander limb inhibits regeneration.

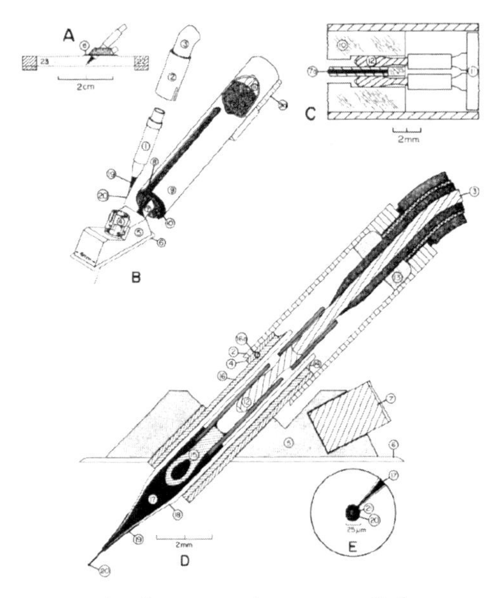

Figure 3.2: Vibrating probe for measuring extracellular currents [40]

Illingworth and Barker later used this device [43] to measure electrical currents emerging during the regeneration of amputated finger tips in children. They observed similar pattern of currents in both amputated limb of salamander and amputated fingers of children. An average current of 22 $\mu A/cm^2$ was found at the amputated finger tips.

Erikson and Nuccitelli [44] were one of the first to hypothesize that electric currents produced within the embryo might guide the movement of migratory embryonic cells. They reported that the motility of fibroblast cells is strongly influenced by DC electric fields. In their experiments, the cells migrated towards the cathode by extending their lamellipodia in that direction. The corresponding *EFS* was between 1 and 10 mV/mm. They also observed that the cells oriented their long axes perpendicular to the field lines. Two different set-ups were used to study this effect. In figure 3.3a cells were observed in a standard chamber which was used for most of their electrotaxis experiments. In figure 3.3b, the standard chamber was modified to study the effect of cross current fluid flow experiments. The results showed that the electrotactic response occurred in a protein-free saline or with a cross-current flow, the field exerts a direct influence on the cells themselves rather than influencing any external factors such as protein gradients as proposed before. They also proposed that the plasma membrane is the most likely target that is influenced by the electric field. As a result of the electric field, a cell experiences half of the voltage facing each electrode. So depolarization at the cathodal end and hyper-polarization at the anodal end equal.

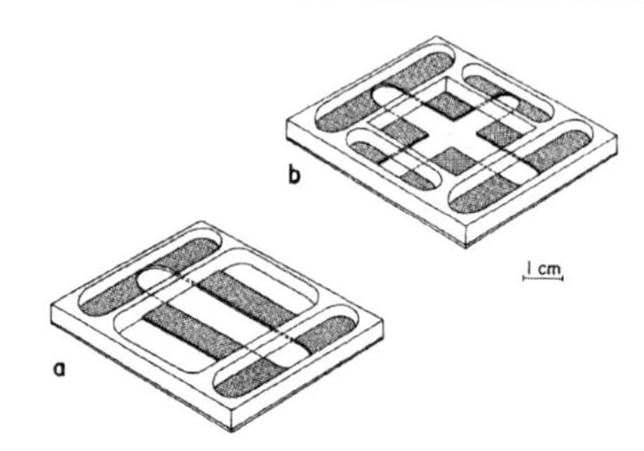

Figure 3.3: Device to study electrotactic behavior of cells [44]: (a) standard electrotactic chamber, (b) chamber with cross-current flow

Erikson and Nuccitelli [44] hypothesized that the EF will induce local membrane potential perturbations that influence the orientation of the integral membrane proteins like ion channels and pumps.

Rapp et al. [45] studied the electrotaxis of granulocytes (Figure 3.4). Granulocytes are cells that are attracted by the sites of inflammation to combat invading microorganisms. The migration mechanism of these granulocytes to the specific site could be explained by both the chemotaxis and the electrotaxis phenomenon. The granulocytes migrate towards the anode and the electrotactic response of granulocytes is a non-cooperative process. The protein essential for electrotactic response is a G-protein [12]. The set-up in figure 3.4 overcomes two problems associated with previous electrotaxis devices:

- A voltage drop occurs at the stimulating electrodes which makes it very difficult to measure the electric field in the chamber
- Products generated at the electrodes diffuse through the medium and affect the medium of cells which can alter the pH value of the solution.

The first problem was solved by using four electrodes, two as power supply electrodes and two as measuring electrodes. The whole device is sealed in paraffin wax to provide electrical insulation. They had to have a short measuring time (<10 mins) before the products entered the chamber.

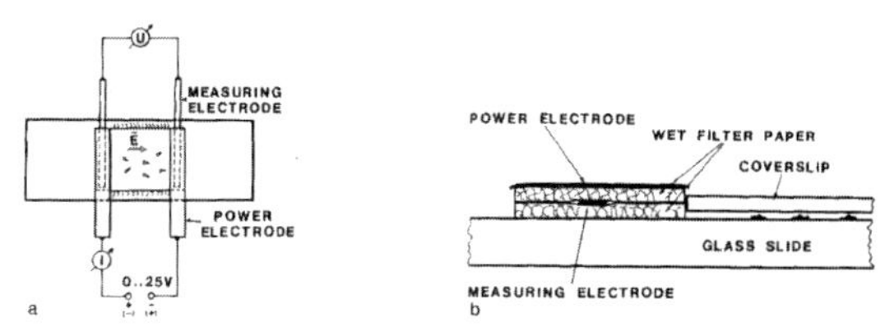

Figure 3.4: Electrotactic device to study and measure electric fields in granulocytes [45]: (a) top and (b) side view of the device.

Rajnicek et al. [46] showed that the rat hippocampal neurons had directional growth in response to the dc electric field in the primary culture. The neuritis oriented perpendicular to the direction of the electric field. Three mechanisms were proposed by which this perpendicular orientation in cells was possible:

- field-induced alignment of the intracellular elements.
- selective retraction of anode facing neuritis
- combination of both these processes.

Nuccitelli et al. [47] showed that extracellular calcium levels influence the electrotaxis behavior of neural cell crests. These cells responded to an *EFS* of 7 mV/mm and migrated towards the negative pole. They studied the effect of calcium ions on electrotaxis of neural crest cells. The device as described in figure 3.3 was used in this study as well. The directed migration was completely blocked when the Ca^{2+} concentration was left unmodified and other ions are added to compete or block the entry of Ca^{2+} ions. They also found that by completely removing the Ca^{2+} ions from the medium, the neural crest cells migrated towards the positive pole. They concluded that gradients of concentration of intracellular Ca^{2+} are involved in signal transduction process. The electric field could increase the permeability of Ca^{2+} ions on the cathode-facing end of the cell.

In 1843 DuBois-Reymond had shown that the human body produces endogenous electric fields naturally. Baker et al. [48] measured the lateral electric fields near the skin wounds of guinea pigs to be 100-200 mV/mm. Nishimura et al. [49] found that human keratinocytes migrate towards the negative pole in *dcEF* whose magnitudes were comparable to those present in wounds of mammals. They observed electrotaxis in keratinocytes for *EFS*s of 100 mV/mm. To completely reepithelialize a wound, keratinocytes wound migrate in a directed, rather than random, fashion. One of the earliest signals to guide directed motility in skin wounds could be electric fields generated at the wound location. Since the keratinocytes show electrotaxis behavior in the physiological range they were able to conclude that *dcEF*s could be the earliest clues for these cells involved in the wound healing process.

Sheridan et al. [50] investigated the influence of the substrate on electrotaxis. Human keratinocytes were cultured (which help during wound healing process) on three different substrates, namely collagen, fibronectin and tissue culture plastics. A physiological electric field of 100 mV/mm is required to direct the keratinocytes migration towards the healing wound. The migratory speed was not influenced by the substrate but only the directionality and cell translocation were affected. A significant difference in the directionality and the average cell translocation was observed. The greatest cathodal migration in response to *dcEFs* was observed for collagen and plastics. Fibronectin surfaces showed a lower directional migratory response.

The work of Erikson and Nucciltelli [44] paved the way for understanding the relation between bioelectric currents inside our body and the electrotaxis phenomenon. Their work brought us one step closer to understanding the electrotaxis phenomenon. Many different species and cells were studied which also showed electrotaxis phenomenon. By the end of 1970s, the knowledge in biochemistry greatly helped to understand various processes leading to electrotaxis. The signaling pathways and mechanisms happening inside the cell were starting to be explored. The significant findings in this period of time laid the foundation for the next era where electrotaxis and wound healing were found to be related.

3.4 1980-2006

By 1990, 14 cell types [51] were identified to show electrotactic behavior. Most of the work after 1995 was focused on understanding the involvement of electrotaxis in the wound healing process. Karba et al. [52] published their work on using *dcEF* stimulation for chronic wound healing enhancement. Their clinical study involved patients who had spinal cord injuries and were suffering from pressure ulcers. Their study showed that the therapeutic effect of electrical stimulation depended on the position and shape of the electrodes used. Healing of the pressure ulcers was significantly enhanced by direct current. They observed highly significant rates of wound healing using such stimulating electrodes which concluded that endogenous electric fields in the skin are just not side effects but play an important role in wound healing processes.

Martin [53] in his work suggested the use of electric fields to regenerate wounded skin perfectly. The wound healing process in embryos can be used as a model to understand this regeneration process since the wound healing in embryos is fast and efficient and results in mostly perfect regeneration of lost tissues. In his review he mentions that human body loses the ability to regenerate tissues after a certain age. By using electric fields these tissues can be stimulated to continue the regeneration process for a longer time. He concluded that-by understanding the wound healing mechanisms in embryos-we could help the clinicians to deal with the skin wound healing better.

Gardner et al. [54] conducted a meta-analysis on the effect of electrical stimulation on chronic wound healing process. They found that the rate of healing per week was 22% for electrical stimulation samples compared to 9% for control samples without electric stimulation. Their study also showed that electrical stimulation was most effective for pressure ulcers.

Zhao et al. [55] reported the electrotaxis behavior of bovine corneal epithelial cells in physiological electric fields. These cells respond rapidly to injury by changing shape and migrating to cover the wound and restore the barrier function. The device used in this study is shown in figure 3.5. Ag/AgCl electrodes were used to generate electric fields in the cell culture chamber. These electrodes were separated from the cell culture medium by use of agar bridges. Cells cultured in 10% foetal bovine serum showed electrotaxis behavior by reorienting perpendicular to the electric field lines for *EFSs* of less than 100 mV/mm. When these cells were cultured in serum-free medium they did not show any electrotactic behavior. When growth factors like epidermal growth factor (EGF), basic fibroblast growth factor (bfGF) and transforming growth factor-beta 1 (TGF-β1) were added the electrotactic behavior in the cells were restored. They found that the directedness of the cell migration was serum-dependent. Highest directedness was observed for a combination of EGF and TGF-β1 serum. Corneal epithelial cells responded to *EFSs* of 100 mV/mm and migrated towards the cathode. Increasing the field strength increased the migration rate. They concluded that therapies involving both growth factors and electric fields are most effective for wound healing.

Zhao et al. [56] also studied the effect of electric fields on bovine corneal epithelial cell sheets. Small *EFSs* of 100 to 250 mV/mm were applied to determine how electric fields interact with other environmental factors in directing these corneal epithelial cell sheets migrations. The cell sheets migrated towards the cathode. The directional migration was voltage-dependent and in low field strengths (200 mV/mm) is required serum in the medium. The cell sheets migrated at a rate of 15 μ/h at 150 mV/mm. Abundant lamellipodia was observed at the leading edges of migrating sheets. Their conclusions were that endogenous electric fields generated by wounded cornea could play an important role in stimulating other environmental factors to promote the change in shape and directed migration of the corneal epithelial cell sheets.

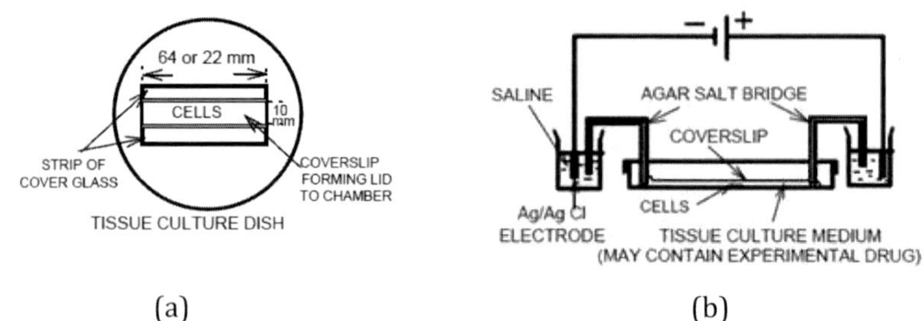

Figure 3.5: Electrotaxis device setup used for studying long term cell migration [55] (a) top view of the device (b) experimental setup of the device

Both early studies by Zhao et al. [55, 56] dealt with studying the electrotaxis properties of bovine corneal epithelial cells. In 1997 they studied the effect of electric field on human corneal epithelial cells (HCEC) as well [57]. HCEC reoriented and migrated towards the cathode. Both the reorientation and the migration were voltage- and serum-dependent. HCEC showed significant orientation and migration behavior for *EFSs* of 150 mV/mm which is close to the physiological *EFS*. From these results they concluded that the endogenous electric fields play an important role in facilitating the shape change and directed migration of corneal epithelial cells during the wound healing process.

Farboud et al. [58] also studies the influence of *dcEFs* on HCEC. They also reported the electrotactic response of HCEC on shape change and migration towards cathode at physiological *EFSs* of 100 mV/mm. The migratory speed and translocation distance of HCEC was similar to keratinocytes. However, corneal epithelial cells demonstrate a more rapid directional response to electric fields compared to keratinocytes. They concluded that endogenous electric fields found in cornea play an important role in human corneal wound healing processes. The endogenous electric field induces orientation and directional response to migratory cells to help in efficiently re-epithealize the wounded area.

Korohoda et al. [59] studied the electrotaxis behavior of amoeba proteus using computer-aided image analysis. The amoeba continued translocation towards the cathode for hours unlike previously reported by Sayers et al. [60]. Korohoda et al. [59] also found that increasing *EFS* in the range of 300 to 600 mV/mm caused the straitening of the cell trajectories and a decreased frequency of the lateral pseudopods formation. The formation of pseudopodia was inhibited on the anode-facing side of the cell. When Ca^{2+} ions were replaced with Mg^{2+} ions in the culture medium they observed a significant reduction in the electrotaxis response of the cells. The localization and kinetics of the primary cell responses occurred at the anode-facing side of the cell. When the polarity of the electrodes was reversed, the cells were able to respond by localization. These cells were studied in the device shown in figure 3.6. Direct current was applied for 180 min using Ag/AgCl electrodes which were immersed in saline-filled beakers and which were connected to the electrotaxis chamber by agar bridges. Amoeba was introduced into the chamber using a micropipette and sealed by silicon grease. Their hypothesis on electrotaxis was much different to that proposed by Jeffe [40]. Jaffe put forward the concept that electrophoresis or electroosmosis of the plasma membrane proteins along the surface of the cells exposed to *dcEFs* is the reason for polarization of the cell. Korohoda et al. suggested that the fast electrotactic responses in amoeba could be explained by the lateral electrophoresis of the ions and the modification of ion gradients near the ion channels.

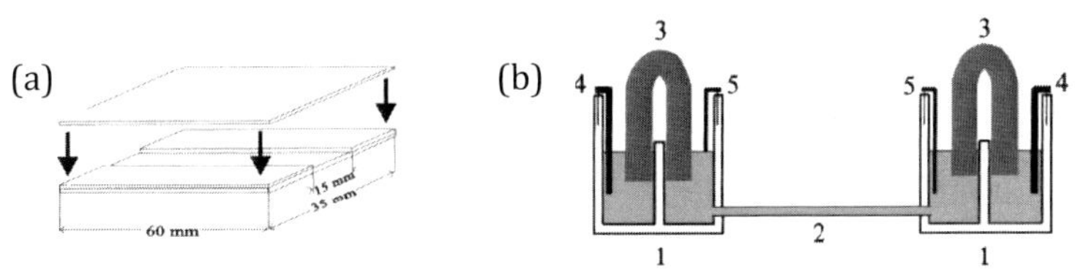

Figure 3.6: Electrotaxis chamber to study electrotaxis behavior of amoeba proteus [59]: a) chamber constructed of cover glass, b) complete plexiglass apparatus.

Puller et al. [61] proposed that the cyclic AMP-dependent protein Kinase A (PKA) plays a role in directed cell migration in human keratinocytes under the influence of *dcEF*. To investigate the role of PKA in electrotaxis they inhibited PKA by an inhibitor which resulted in 53% reduction in directional response of keratinocytes while not affecting the general cell motility. They also investigated if protein kinase C (PKC) would be involved in the signaling pathway. By using a PKC inhibitor like D-erythro-sphingosine they saw no effect on the directional response of the cells. Their work concluded that divergent kinase signaling pathways are involved in general cell motility and the role of PKA is involved in regulating directional response of cells to electric fields.

Djamgoz et al. [62] studied the involvement of voltage-gated Na^+ channels on electrotaxis behavior of rat prostate cancer cells. The prostate cancer cells show electrotactic behavior and Na^+ channels are involved in their electrotaxis of cells. They studied two different cell lines derived from the same prostate tumor differing in their metastatic ability. *dcEF* strengths in the physiological range of 0.01-0.4 V/mm were applied using the same electrotaxis device shown in figure 3.6. Tertrodosin and veratridine were used to block the Na^+ channel and to open the Na^+ channel, respectively. The highly metastatic MAT-LyLu cells (prostate cancer cells) responded to *EFSs* by migrating towards the cathode. They also observed that the weakly metastatic AT-2 cells did not respond to the *EFSs*. Tetrodoxin inhibited and veratridine promoted electrotaxis. They were able to show that electrotaxis behavior could be controlled by Na^+ channels.

Fang et al. [63] investigated the role of growth factors and extracellular calcium on electrotaxis behavior of human keratinocytes. After *EFS* of 100 mV/mm, human keratinocytes migrated towards the cathode at a rate of 1 μ/min. In the absence of growth factors, the migration speed decreased but directionality was maintained. Epidermal growth factors alone restored the cell migration rates significantly. The effect of extracellular calcium on the migration behavior of human keratinocytes was also studied. Changes in calcium concentration did not affect the electrotaxis behavior significantly but addition of chelator ethyleneglycol-bis (β-aminoethylether) to migration medium inhibited keratinocytes migration ability. Trollinger et al. [64] followed up this work and showed that Ca^{2+} channels are required for electrotaxis in human keratinocytes. They investigated the role of calcium influx on the directionality and migration speed of human keratinocytes using Ca^{2+} channel blockers. Calcium channel blockers like Verapamil, amiloride, inorganic Ca^{2+} and Ca^{2+}-substitute Sr^{2+} blockers were used to study the electrotactic behavior of these cells. Verapamil and amiloride had no effect on the elextrotaxis behavior whereas Sr^{2+} strongly inhibited the electrotaxis behavior. They concluded that Ca^{2+} channels are involved in electrotaxis.

Wang et al. [65] used a lens epithelium cell monolayer as a model to show that electric fields and mitogen activated protein kinase (MAP kinase) signaling can regulate wound healing process. They used a MAP kinase inhibitor to study the signaling pathway involved in lens epithelium cells. They also found that electric fields enhanced the active extracellular signal-regulated kinase (ERK 1/2) and the inhibitor inhibited ERK 1/2. They reported that the inhibitor deceased the wound healing process with or without the application of electric fields. They concluded from their study, that the ERK signaling pathways were involved in the wound healing process of lens epithelium cell monolayers.

Mycielska and Djamgoz [66] in 2004 explained the cellular mechanisms of the electrotaxis phenomenon. They postulated that the passive influx of Ca^{2+} on the anodal site increase the local intracellular Ca^{2+} ion concentration whereas passive efflux decreases the intracellular Ca^{2+} ion concentration on the cathodal site. These changes give rise to a push-pull mechanism causing the movement of the cell towards the cathode. Ca^{2+} ion channels, Na^+ ion channels, protein kinases, surface charges, growth factors and electrophoresis of proteins are also involved in electrotaxis. The electrotactic responses of strongly metastatic prostate and breast cancer cells are much more prominent, and the cells move in the opposite direction compared with corresponding weakly metastatic cells. This could have an important implication for clinical purposes. They concluded that electrotaxis plays an important role in both cellular physiology and pathophysiology.

Sun et al. [67] studied the migration of human fibroblasts in 3-dimensional collagen gel. They observed that 3-d migration involves both the *EFS* as well as the collagen concentration. The cells responded to an *EFS* as low as 0.1 V/cm in 3-d unlike in 2-d. They concluded that the ability to apply electric stimuli and to control 3-d cell migration would provide an alternative methodology for regulation of engineered tissue contructs.

Reid et al. [68] investigated the role of electric currents during wound healing in rat cornea and studied the spatial and temporal distribution of endogenous electric fields in rat corneal wounds. They found that Cl^- and Na^+ are the major components of that electric currents. Na^+ is the major component of ionic transport in the resting wound rat cornea and wound center leakage current. Cl^- plays a major role in the endogenous electric current at wound edges. Enhancing or decreasing Cl^- flow increased or decreased the endogenous currents, respectively. These changes in wound current directly affected the rate of wound healing in vivo. They suggested that this result could have therapeutic potential in wound healing processes.

Ogawa et al. [69] proposed a physical model for electrotaxis of paramecium cell. They derived a mathematical model for the torque that orients the cell towards the cathode which is a characteristic ciliary motion observed in paramecium cells called the Ludloff phenomenon. Their numerical simulations confirmed realistic behavior like U-turns of cells.

Funk and Monsees [7] studied the effects of electromagnetic fields on cells. They discussed the physiological and therapeutical approaches and molecular mechanism of interactions in cells. Their study focused on electrotaxis experiments and application this knowledge to electromagnetic fields-based therapy for wound healing and bone fractures. They discussed the contact guidance phenomenon first proposed by Brunette in 1986 [70]. Contact guidance can be defined as the process of adherent growing cells to migrate and align in the direction of the surface structure. They tested this by patterning the surface using titanium oxide lines of 12 nm in height, 1 mm in length and 20 μm in width. Most of the osteosarcoma cells aligned in the direction of the lines. Due to the physicochemical properties of titanium oxide surface, other mechanisms may be involved to influence the focal adhesion contacts and cell guidance. One such factor could be the gradients of the electrostatic potential at the titanium-oxide interface. This study also included the therapeutic relevance of electromagnetic fields. The application of electromagnetic fields to stimulate osteogenesis is based on the idea of stimulating the natural endogenous potential of the bone. The most effective device for healing bone fractures was found to be the devices that used time-varying or pulsed electromagnetic fields (1-100 Hz) inducing electric fields at the fracture site. Electromagnetic fields are only effective at extremely low frequencies (8 60 Hz) and low amplitudes (< 1Gs).

In 2006, Zhao et al. [4] found the genes responsible for electrotaxis phenomena in cells. They showed that electric fields of physiological strength direct cell migration during wound healing. The cells use this electric field as a prime cue for migration. When this endogenous electric field is manipulated, it affected the wound healing process in vivo. The electric stimulation activates Src and inositol-phospholipid signaling which polarizes in the direction of migration. They genetically modified phosphatidylinositol-3-OH kinase-γ (PI(3)Kγ) and found that this decreased the electric field-induced signaling and inhibits the directed movement of cells in an electric field. They also found that deletion of the tumor suppressors phosphatase and tensin homolog (PTEN) enhanced the signaling and electrotactic response. They thus concluded that the genes responsible for electrotaxis phenomenon in cells were PI(3)Kγ and PTEN.

The years 1980-2005 answered many unresolved questions in the field of electrotaxis and contributed to the understanding of this phenomenon during this time. Works of Djamgoz et al. [62] and Wang et al. [65] showed the importance of Na^+ and Ca^{2+} ion channels and their involvement in electrotaxis phenomenon. The various signaling pathways inside the cells responsible for electrotaxis phenomenon were also studied. Many studies were focused on using this property of cells to improve clinical therapies.

3.5 2006-2010

During the last years the focus in the field of electrotaxis shifted from cell biology studies to engineering relevant devices to study this phenomenon. Nanotechnology coupled with micro-fluidics was used to develop lab-on-chip type devices to study cells. This part of the review will focus especially on these devices used for studying electrotaxis.

Chao et al. [71] studied the effect of *dcEFS* on ligament fibroblast migration and the wound healing process. They applied static and pulsing *dcEF* to calf anterior cruciate ligament (ACL) fibroblast cells. They tested the motility of these ACL cells on type I collagen substrates. The type I collagen expression increased after exposure to electric fields. The applied electric fields augment ACL fibroblast cell migration and biosynthesis and provide a good proof for the electric fields to be used for enhancing the ligament healing and repair.

Finkelstein et al. [72] showed that the polarization of charged cell surface proteins induced by electric fields are not determined by the direction of electrotaxis. They conducted cell migration studies on mammalian cells (3T3, HeLa and CHO cells). The hypothesis that an electric field polarizes the charged cell surface molecules and these polarized molecules drive directional motility was tested. The device used in this study was the same as used by Chao et al. [71]. It was known that sialic acid, which is negatively charged, contributes to the bulk of the surface charge in the cell and redistributes preferentially to the surface facing the direction of motility. They treated cells with neuraminidase to remove sialic acids and obtained cells with less surface charge. The neuraminidase inhibited the electric field-induced directional polarization while avidin treatment with its positive charge reversed the directional polarization of sialic acids. They also found that neuraminidase treatment inhibited directionality but did not alter the speed of migration whereas avidin treatment had no significant effect on the speed and directionality of migration. They concluded that the polarization of the charged cell surface proteins is neither necessary nor sufficient to cause migration.

Sato et al. [73] studied the input-output relationship in the electrotactic response of dictyostelium cells by means of a device which could reproduce the electrical signals applied to the cells consistently. By this, the migration response could be analyzed quantitatively (Figure 3.7).

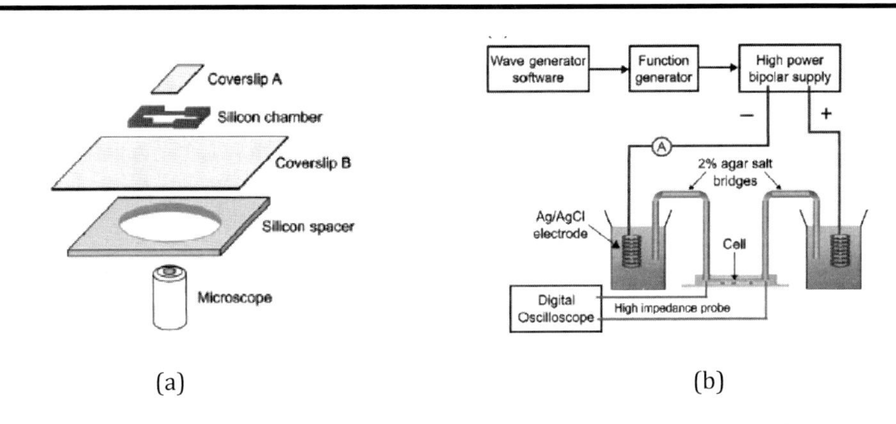

Figure 3.7: Experimental set-up of the device for studying the input-output relation in electrotactic response of cells [73]: (a) cell chamber, (b) experimental set-up

The device consists of a glass cover slip on top of which another cover slip was placed by a silicon spacer. This chamber was connected to the power supply by means of a long agar bridge (15 cm), buffer solutions and Ag/AgCl electrodes. The *EFS* was calculated by measuring the voltage between the two ends of the channel and by dividing this value by the length of the channel. *EFS*s of 0.25 to 10 V/cm was applied in this study. They observed random migration of cells without electric field but upon application of electric field, cell migration speed increased by a factor of 1.3 times and showed directed migration towards the cathode. They found that the direction migration was dependent on the square of the *EFS*.

Song et al. [8] conducted electrotaxis experiments using an electrotaxis device shown in figure 3.8. In order to avoid any chemical gradients generated by electric fields inside the chamber, the flow of medium was allowed in a controlled manner. Their device could be used for both planar cell culture as well as tissue culture. The electrotactic chamber is constructed using glass slides to form a glass well as shown in figure 3.8 a, b. The chamber base is then fabricated by gluing two cover glasses to the petri dish (fig. 3.8 c, d). The cells are cultured in the glass well overnight. The glass well is removed and the top is sealed using another cover glass. The lid of the petri dish has two holes to accommodate the agar bridges. This set-up was also used by Zhao et al. [4] to carry out wound healing experiments.

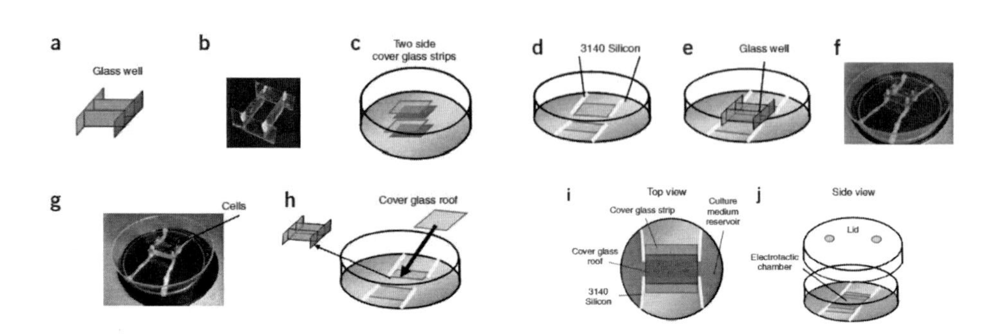

Figure 3.8: Electrotactic cell chamber [8]:(a,b) glass well, (c,d) set-up of the electrotactic chamber base with two isolated reservoirs at the sides of the electrotactic chamber, (e–g) placing the glass well on top of the electrotactic chamber base before and after seeding cells in the well, (h) cover slip roof glued onto the cover glass strips completing the electrotactic chamber, (i) top view of the electrotactic chamber covered with the cover glass roof, (j) electrotactic chamber covered with a lid, ready for EF experiment.

Nuccitelli et al. [74] mapped the electric fields around mouse and human skin wounds using a novel bioelectric field imager (BFI) (figure 3.9). The BFI is a non-invasive technique to measure *EFS*s near a wound. A small metal probe vibrates above the wound to detect the surface potential of the skin through capacitive coupling. They were able to measure *EFS*s of upto 177 mV/mm immediately upon wounding and the electric field lines pointed away from the wound. They found that-since wound currents flow immediately after wounding-this could be the first signal indicating a wound in the skin. They found that the wound current beneath the skin remained in the range of 150-200 mV/mm for about 3 days and then began to decline over the next few days and reached zero once the wound healing process was complete. By using sodium channel blockers like amiloride the wound healing rate slowed down by 64 %, whereas Cl^- channel activators like prostaglandin E2 increased the wound healing rate by 82 %.

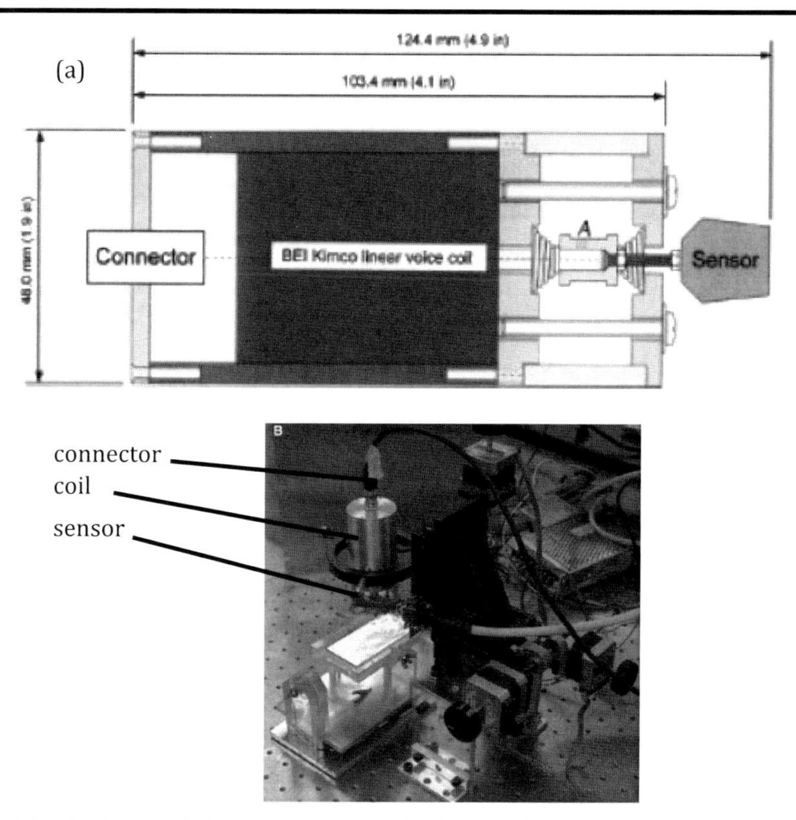

Figure 3.9: Bioelectric field imager to map the electric fields around a wound [74].

Lin et al. [75] studied the electrotaxis phenomenon of lymphocytes in vitro and in vivo. A combination of the modified Transwell assay [76] and a microfluidic device was used in this study to show that lymphocytes migrate towards the cathode at physiologically relevant *EFS*. Figure 3.10 shows the Transwell assay and the microfluidic device used in this study. The channel in the microfluidic device was fabricated using laser cutting of a Melinex non-adhesive sheet. They used two pipette tips as medium reservoirs. The electrical stimulation activates intracellular kinase signaling pathways which were similar to the chemotaxis phenomenon. The GFP-tagged immunocytes were traced in the skin of mouse ears. It was found that motile cutaneous T-cells migrate towards the cathode in an applied *dcEF*. They concluded that lymphocyte positioning within tissues can be manipulated by applying external electric fields and are also influenced by endogenous electric field gradients. This could allow novel approaches to manipulate the positioning of lymphocytes to improve vaccine and antitumor therapies.

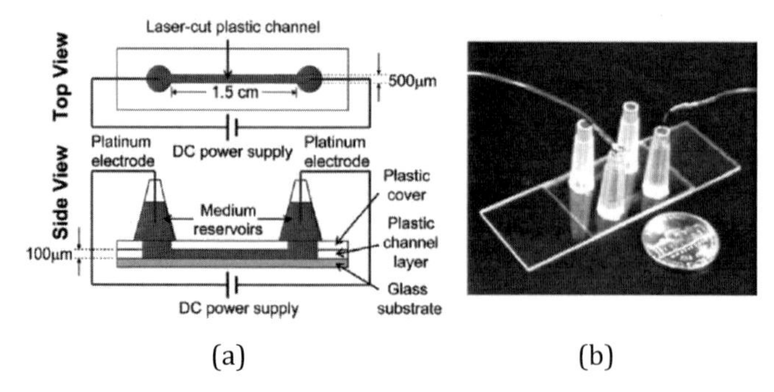

(a) (b)

Figure 3.10: Combination of transwell assay and microfluidic device for studying electrotactic behavior of lymphocytes [75]: (a) schematic and (b) photograph of the device

Yang et al. [77] designed a novel electrical bioreactor for studying wound healing (Figure 3.11). It consists of three main components: bottom chamber, cover plate and culture chamber lid. Instead of using agar glass rod bridges as in a conventional set-up, they fabricated two trough-holes, which are filled with agar. These agar troughs connect the electrodes to the cell culture medium thereby slowing down the flow of toxic products from the electrodes into the cell culture medium. Uniform electric fields were generated to study the electrotaxis of Hy926 cells. The electric fields inside the bioreactor were simulated using ANSYS software. The cells were seeded on a separate glass slide and then placed into the bioreactor for application of the electric field. It was shown that the Hy926 cells migrated towards the cathode.

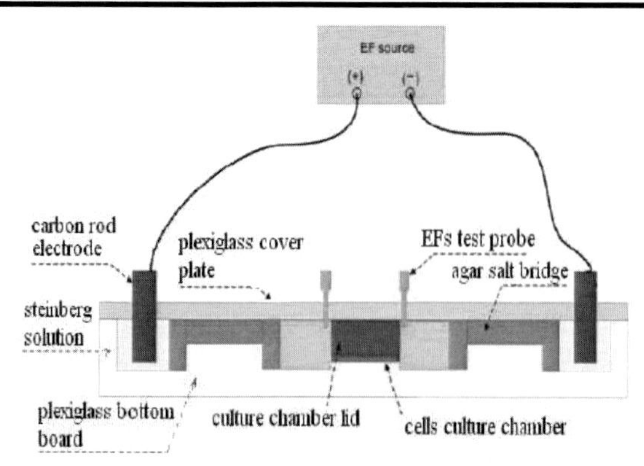

Figure 3.11: Bioreactor to study wound healing [77]

Huang et al. [78] studied the electrotaxis phenomenon of lung cancer cells in a device that showed three areas with different *EFS* (Figure 3.12). The device has three different regions of different thickness to generate the different *EFS* levels. The microfluidic channels were fabricated by laser cutting machine on a polymethylmethacrylate (PMMA) substrate. This substrate was then bonded to blank cell culture slips using a double-sided tape. In figure 3.1a the assembly view of the device is shown. The device consists of a plastic culture slip on top of which the double sided tape is stuck. The other side of the tape has the PMMA sheet with the inlets and outlets for medium and agar bridges.

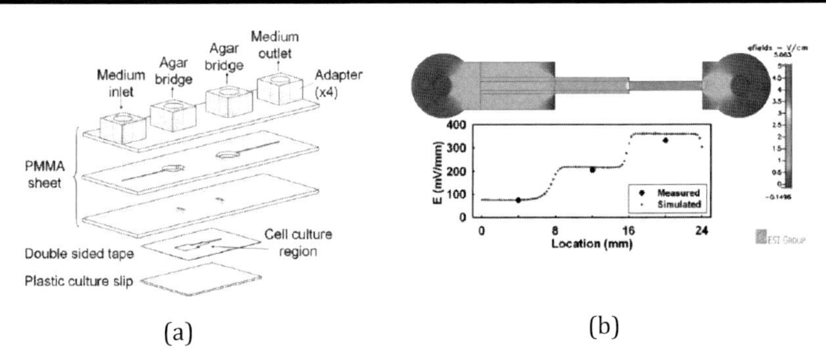

(a) (b)

Figure 3.12: Electrotactic device for studying cancer cells in different electric fields [78]: (a) assembly view, (b) simulated *EFS* distribution inside the device

Lung cancer cell lines respectively were studied with high and weak metastasis, CL1-5 and CL1-0. Both the cell lines had different response to applied *EFS*s in the range of 75 375 mV/mm. CL1-5 cells migrated towards the anode while CL1-0 did not show any response to electric field. They concluded that-based on the cell orientation and responses-the two responses of the cell lines to electric fields involve two different signaling pathways.

Ozkucur et al. [24] studied the influence of the calcium level in osteoblast-like cells. A commercial electrotaxis device was used in their study (fig. 2.3). Regulation of intracellular Ca^{2+} levels plays an important role in *dcEF* influence on cells. At strong *dcEF* strengths, the voltage-gated calcium channels are activated and may be responsible for intracellular Ca^{2+} ion kinetics in osteoblast-like cells. They concluded their findings by reporting that the Ca^{2+} activity in osteoblast-like cells could be helpful in understanding both the guided migration and the elongation of the cells in direct current electric fields.

Wang et al. [79] designed a microfluidic device to study the asymmetric cancer cell filopodium growth induced by electric fields. They utilized the structured illumination nano-profilometry (SINAP) technique to quantitatively study the variations of cancer cell filopodia under the influence of *dcEF* (Figure 3.13).

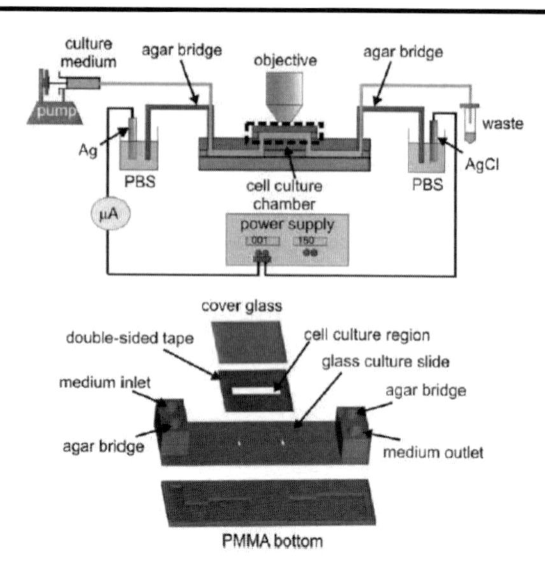

Figure 3.13: Schematic diagram of the microfluidic device used for studying cancer cell filopodium under the influence of electric fields [79]

The principle of SINAP is based on the combination of structured illumination microscopy and differential height measurement. They observed filopodia with diameters of less than 200 nm without florescent labeling. The channels in the microfluidic device were fabricated by laser ablation of PMMA. A double-sided adhesive tape was used to define the cell culture chamber. Four circular holes are used for medium inlet and outlet and agar bridges to transfer current from the buffer solution with electrodes. They reported that, for lung cancer cells with high migration and invasion properties, the growth of filopodia is predominantly on the side facing the cathode. This was especially for *EFSs* in the 180 250 mV/mm range whereas it decreased as the *EFS* was higher than this range.

Rezai et al. [80] designed a microfluidic device to study the electrotaxis property of caenorhabditis elegans (worm). They were able to control the movements of these worms using electrical stimulus in a microfluidic environment (Figure 3.14).

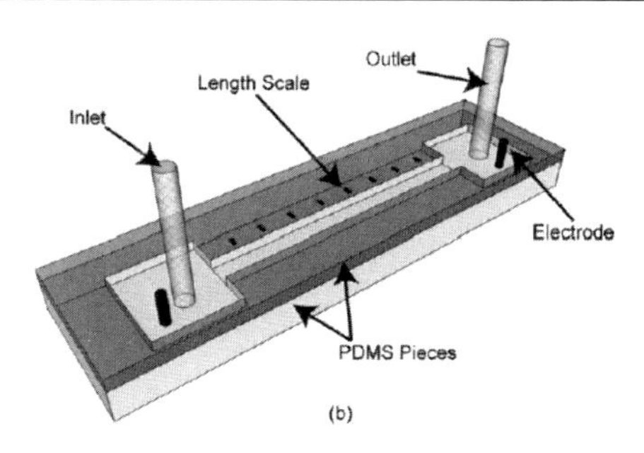

Figure 3.14: Micro-fluidic device to study the electrotactic behavior of worms [80]

The electrotaxis device consists of four major components: a microchannel with electrodes at each end, inlet and outlet tubes to insert worms into the device, a power supply unit with electrodes and a microscope for live imaging. The worms were pumped into the microchannel using a syringe pump till they were at the middle of the channel. The pump was then disconnected and the medium in the reservoirs were leveled. Current was applied using the electrodes and the worms were observed under a microscope. The results revealed that electrotaxis is mediated by neuronal activity that varies with age and size of the animals studied. The speed of migration was not affected by the *EFS* rather each developmental stage responds to a specific range of *EFS* with a specific speed.

Li et al. [81] investigated the electrotaxis property of activated T-lymphocytes using a microfluidic device. They used two set-ups to study these cells under the influence of electric fields (Figure 3.15). The channels in the electrotaxis device of figure 3.15a were fabricated using soft lithography with PDMS. The master for the PDMS mold was first made on a silicon wafer and afterwards transferred to PDMS. Two holes were punched at the end of the channel and platinum wires were inserted to act as the electrodes. The PDMS mold and glass slide were sealed using plasma bonding. The device shown in figure 3.15b was fabricated using photolithography technique. Two different masks for channels and electrodes were used. The channels were etched into the glass slide using wet isotropic etching and the electrodes were created by powder blasting technique. T-lymphocytes migrated towards the cathode in a *dcEF*. Using these two microfluidic devices they were able to show that anti-CD3/CD-28 antibody activated human blood T cells migrated towards the cathode in a *dcEF*.

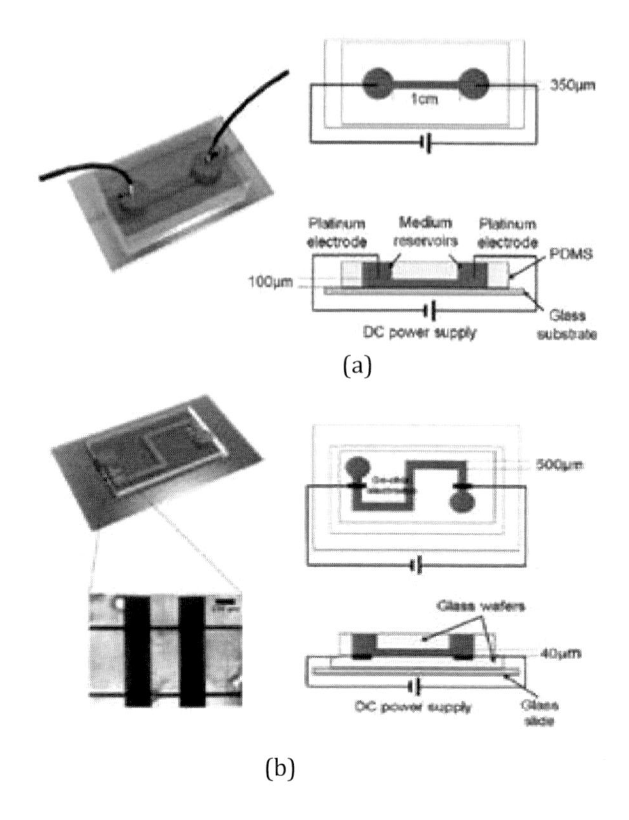

Figure 3.15: Microfluidic setup for studying electrotaxis behavior of T lymphocytes [81] (explanation in text)

Maniere et al. [82] used the electrotactic property of C. elegans cells for sorting them. Electric fields were applied in a micro-fluidic device to sort the cells based on their crawling speed (figure 3.16). They were able to quantitatively measure the effect of mutations and aging by relating it to the crawling velocity of worms. An agar gel was cast by using a PDMS block in the center. A second layer of agar was added around the PDMS. Once solidified, the PDMS block was removed resulting in an agar pad which was placed in an electrophoresis device. The authors were able to show that worms with different locomotory phenotypes can be sorted spatially where fast moving worms move away from the slow moving ones. They concluded by stating that C. elegens can be used as a model for understanding neurodegenerative diseases and using the electrotaxis phenomenon for sorting could be beneficial in therapeutic bio-molecules.

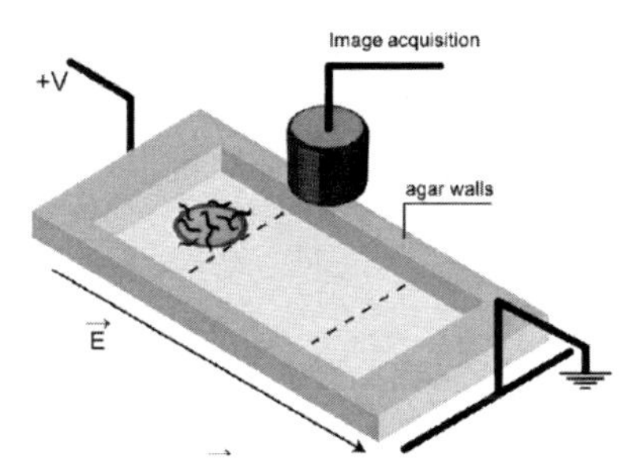

Figure 3.16: Electrophoresis device to sort worms using electric fields [82]

The application of micro-fluidics and silicon micro-technology greatly helped in developing new devices to study the electrotaxis property of cells. Various technologies like soft-lithography, photolithography and laser cutting were utilized to fabricate these devices. Using these new devices, it became possible to study the effect of *EFS* on cells.

3.6 2010-present

The future direction of electrotaxis and its possible application is discussed in this section. All the previous work has greatly contributed to the understanding of how electrotaxis works in cells and this knowledge would be used for studying stem cells and wound healing therapies.

One of the first published works on use of electric fields to differentiate a stem cell into a specific type of cell was demonstrated by Serena et al. [7]. Although this work was done earlier it is relevant in this section. They showed that stem cells could be differentiated into cardiac cells by using *dcEF*. Custom-built bioreactor was used to stimulate stem cells to differentiate into cardiac cells along with the production of reactive oxygen species (Figure 3.16).

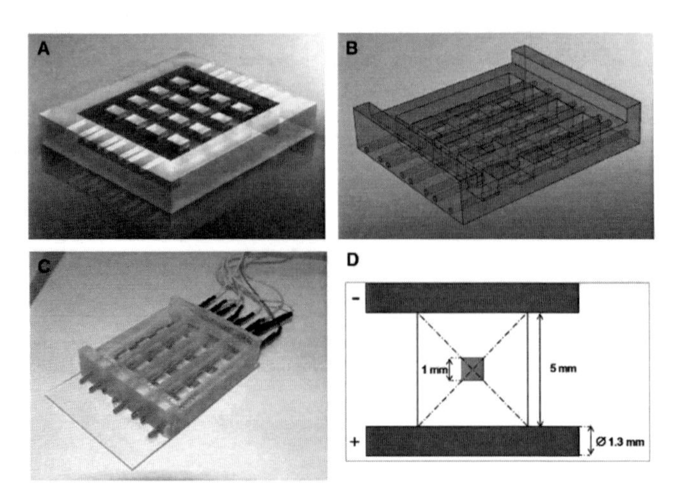

Figure 3.16: Bioreactor to study stem cell differentiation by electrical stimulation [7]

The effect of electrode material (stainless steel, titanium nitride-coated titanium, and titanium), length of stimulus (1 to 90 s) and age of embryoid bodies at the beginning of electrical stimulation (4 and 8 days) with respect to the reactive oxygen species generation were investigated. The highest concentration of reactive oxygen species was generated with a stainless steel electrode stimulated for 90 s and for 4 day-old embryoid bodies. Cardiac differentiation was evident from the spontaneous contractions and expression of troponin T. The authors concluded from these results that electrical stimulation plays an important role in cardiac differentiation of stem cells through the mechanism of intracellular generation of reactive oxygen species.

Zhao et al. [83] investigated the electrotaxis behavior of bone marrow stem cells. These cells migrated towards the anode in *EFS* of 25 mV/mm. Maximum migration speed of 42 μm/hr at 300 mV/mm was observed. Electric fields play an important role in the early-directed migration of bone marrow stem cells, which can be controlled or enhanced by *EFS*.

Feng et al. [84] reported the electrotaxis behavior exhibited by neural stem cells. The cells responded to small *EFS* of 16 mV/mm and migrated towards the cathode. The migration directedness and distance increased with the increase in *EFS*. The Rho-kinase inhibitor did not affect the directionality of the stem cells in electric fields. Blocking the cytokine receptor did not influence the electrotaxis behavior which normally inhibits the chemotaxis behavior in the same cells. The authors concluded that applied electric fields could be potentially exploited to guide stem cells to injured sites in the brain to improve the outcome of the various diseases.

3.7 Conclusions

From the time it was first observed in 1890's to the present time electrotaxis behavior of cells was studied by a lot of researchers. We now have a good though not complete understanding of this behavior of cells. Some of the notable discoveries in this field were measuring the electric fields around a wound, identifying the genes responsible for this behavior and differentiating a stem cell using electric fields.
The application of micro-fluidics greatly helped the cell migration research. The advantages of using micro-fluidic devices in cell migration are:

- stable and defined chemical concentration gradients
- miniaturization
- low reagent consumption
- high throughput experimentation.
- the electric fields can be easily manipulated and controlled in such devices.

There are still some hurdles we have to overcome before the micro-fluidic devices can be standardized in all the labs for cell migration studies. These limitations include:

- standardization of the materials and processes to fabricate these devices
- a standardized and user friendly micro-fluidic platform could be developed for use in different experiments.

The future of this field is directed towards application of the electrotaxis effect in stem cell research and clinical therapies for injuries. Moreover, response to cells in an *EFS* gradient would help us to understand this effect in a different point of view. This study is presented in the future chapters of this thesis.

4 MATERIALS AND METHODS

This chapter introduces the materials and their properties used for fabricating an electrotaxis device for application with *EFS* gradients. The main material used in the fabrication process involves PDMS. The chemical and physical properties of PDMS are explained. The fabrication process of the electrotaxis device is also described in detail. The cell culture procedures used in this work and also the different configurations of microscopes are explained. Lastly, it is shown how the migration data collected is analyzed.

4.1 Polydimethylsiloxane

Polydimenthysiloxane (PDMS) is one of the most widely used polymer for fabricating biological devices because it is non-toxic to cells (bio-compatible) and impermeable to liquids but permeable to gases. PDMS is optically transparent, elastic and inexpensive. Other major advantages of PDMS over other polymers are the ease of fabrication and bonding to other material surfaces. PDMS is also a good choice of material for rapid prototyping of devices.

The chemical structure of PDMS is shown in Figure 4.1. The chemical formula of PDMS is $CH_3[Si(CH_3)_2O]_nSi(CH_3)_3$ where n is the repeating monomer $[SiO(CH_3)_2]$. Depending on the size of the monomer unit the non-cross-linked PDMS could be liquid (low value of n) or semi solid (large value of n). The siloxane bond allows for a flexible polymer chain and a high level of viscoelasticity could be obtained.

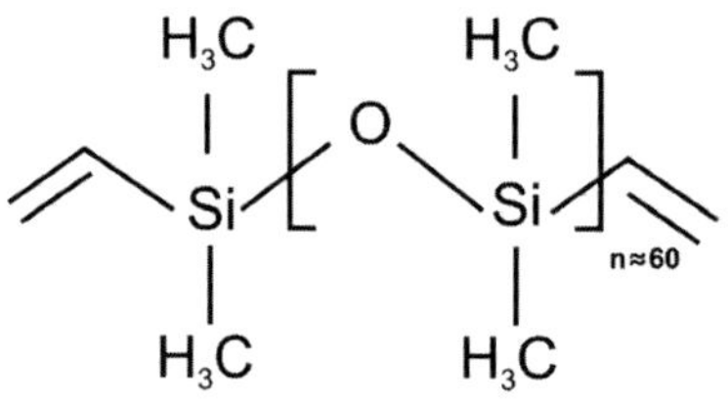

Figure 4.1: Chemical structure of PDMS

For this work the PDMS type Sylgard 184 (Dow Corning Inc.) was used. The silicone elastomer kit consists of the base PDMS and a cross-linking agent (silicone resin solution). PDMS is transformed into a 3-dimensional network via cross linking by an organometalic cross linking reaction. The base PDMS polymers contain vinyl groups. The cross linking oligomers has a minimum of three silicon hydride (SiH) bonds and a platinum based catalyst solution that catalyzes the addition of SiH bonds across the vinyl groups forming $Si\text{-}CH_2\text{-}CH_2\text{-}Si$ linkages. The multiple sites on both the base PDMS and cross-linker allows for the formation of a 3-dimensional network. The ratio of cross-linker and base PDMS decides the stiffness of the cured elastomer. The stiffness is directly proportional to the amount of cross-linker in the elastomer. This polymerization reaction is accelerated by proving heat.

Standardized procedures to fabricate microfluidic devices and molds are reviewed by Duffy et al. [85], McDonald et al. [86] and Ng et al. [87]. They comprise of the following steps:

- Design of the microstructure by computer-aided-design (CAD) tools

- Transfer design onto a mask. A transparency sheet or mask plate is fabricated by printing or by using laser lithography depending on the size of the structures. This is later used as a photomask for ultraviolet (UV) photolithography.
- Photoresist spin-coating. A positive or negative photoresist depending on the requirement of the structures is spin-coated on a clean silicon wafer. The heights of the structures are decided on the thickness of the photoresist film set by the number of revolutions during the spin-coating process.
- UV exposure. The silicon wafer coated with the photoresist film is exposed to UV light through the photo mask. The exposed areas are cross-linked.
- Developing the microstructures. A developing reagent is used to dissolve the unexposed areas of the resist to form the microstructures on the silicon wafer. This silicon wafer would be used as a master structure mold for replicating PDMS molds.
- Surface treatment of silicon wafer. In order to release the PDMS mold from the silicon wafer the surface of the silicon wafer is treated with fluorinated silanes. This treatment makes the silicon surface hydrophobic and prevents PDMS from forming an irreversible bond.
- PDMS preparation and curing. PDMS base is mixed with curing agent in a ratio of 10:1 to form a PDMS liquid elastomer. This PDMS elastomer is poured on top of the silicon wafer and is cured (cross-linked) by heating at 70 °C for 1 hour. This process of replicating the structures using PDMS is called soft lithography.
- PDMS peeling. PDMS is peeled carefully and the required microstructures are formed on the PDMS mold.
- Surface modification of PDMS mold. The PDMS mold is treated in O_2 plasma to make the surface and channels hydrophilic. O_2 plasma also helps PDMS to form a permanent seal with glass or another PDMS surface.

4.2 Device fabrication

This section describes the fabrication process of the microfluidic device used to study electrotaxis behavior of cells in both homogenous and non-homogenous electric fields. The schematic diagram of the electrotaxis device can be seen in Figure 4.3. The set-up is similar to the standard electrotaxis experiment. The device is simple to fabricate and live imaging is possible for up to 6 hours. The difference between a standard electrotaxis device and the device used in this study is the incorporation of two additional electrodes to generate non-homogenous electric field configurations. The two electrodes change the direction of electric field lines and produce complex electric field patterns.

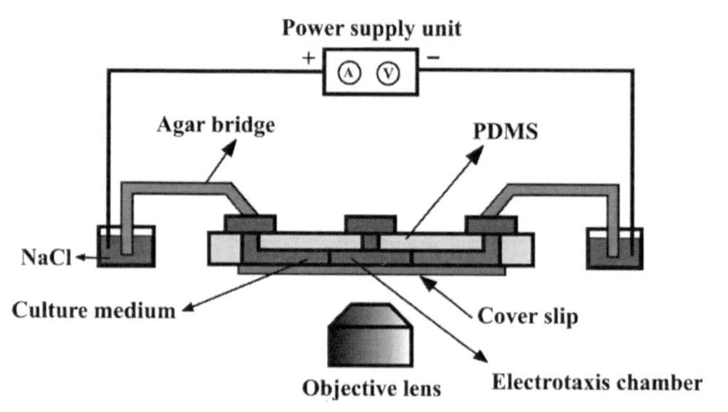

Figure 4.3: Schematic set-up of the electrotaxis device used in this work to investigate cells under the influence of *EFS* gradients

The device is made up of three main components: bottom plate, top plate and electrical power supply unit.

(i) Bottom plate:

The fabrication process of the bottom plate is illustrated in Figure 4.4. The bottom plate consists of a glass cover slip (VWR International, catalog 631-0178) and a PDMS channel in the shape of a cross. The four ends of the cross-shaped channels form the four electrodes. This cross shape thus produces non-homogeneous *EFS* when all the four electrodes are activated. The PDMS channels are produced on top of this cover slip by soft lithography. Firstly, an adhesive (3M 467 MP, 200MP Adhesive) sheet is cut out in the shape of a cross. The thickness of this sheet is 0.6 mm. This structure is then placed on top of a clean cover slip such that it is aligned to the center of the cover slip. PDMS elastomer with a PDMS-to-curing-agent ratio of 10: 1 is then poured on top of the cover slip and around the structure to form a thin layer. After 5...10 minutes, the PDMS takes the shape of the cross-structure forming a thin film on the cover slip. Then the PDMS is cured at 70 °C for 1 hour on a heating plate. After curing, the cross structure is carefully removed to form the channels on the bottom plate.

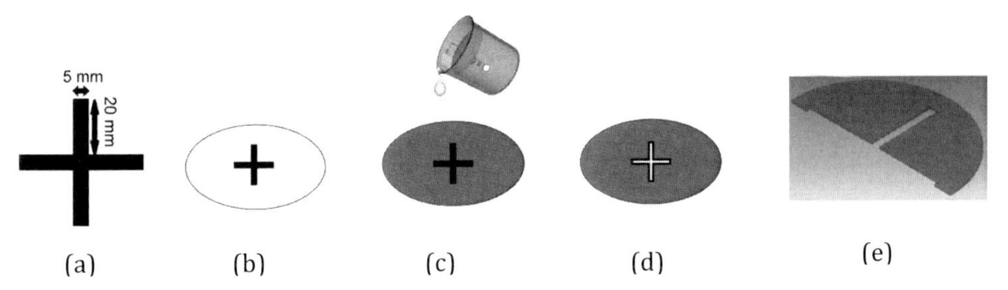

Figure 4.4: Fabrication process of bottom plate: (a) dimensions of cross structure, (b) aligning the cross structure on top of the cover slip, (c) pour PDMS on top of the cover slip, (d) curing of PDMS and removal of the cross structure, (e) cross-section view.

(ii) Top plate:

The top plate is used to seal the channels. Figure 4.5 shows the top plate. This is a commercially available cell culture well (NUNC IVF 4-well dishes, catalog no. 144444). Four holes of 5 mm each are drilled in the center of each chamber which serves as the opening to each channel (as shown in Figure 4.6).

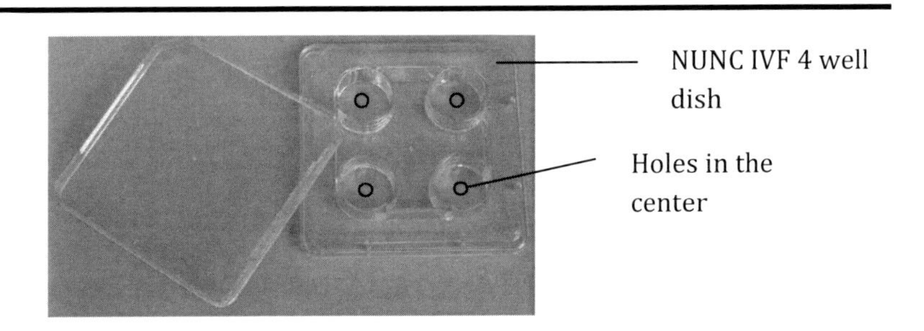

Figure 4.5: Top plate (NUNC IVF 4-well dish) with holes in the center of each dish.

(iii) Electrical power supply unit setup:

The set-up consists of a power supply unit (Amersham Pharmacia Biotech, Freiburg, Germany), buffer solution beakers (0.9 % NaCl solution) with platinum wires (0.2 mm diameter, Agar scientific, Essex, UK), and 20 cm long glass tubes filled with agar (2% Agar in phosphate buffer saline solution). The direct current is transferred from the power supply unit to the buffer solution using the platinum wires. One end of the agar bridge is immersed in the buffer solution and the other end is contacted with the top plate chamber. The top plate chamber is filled with cell culture medium so that the channels are also filled.

4.3 Assembly of the electrotaxis device

The bottom plate is aligned to the top plate such that the four holes in the center of the chambers are above the ends of the channels. A thin layer of PDMS is spread on the bottom side of the top plate and aligned with the bottom plate and cured at 70 °C for 1 hour to form a seal between the two. Assembled view of the electrotaxis device is shown in figure 4.6.

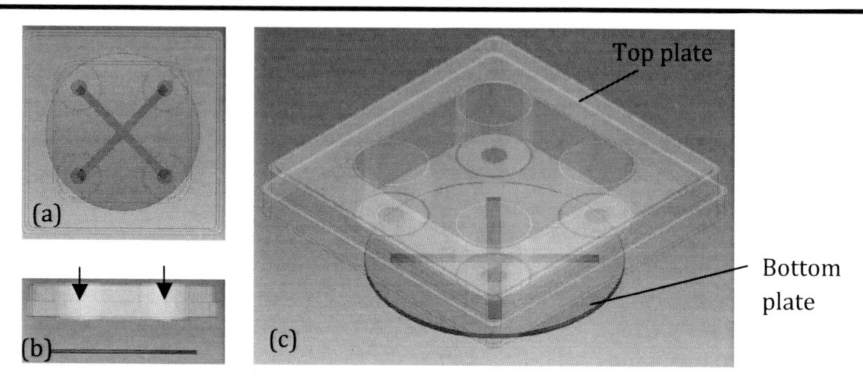

Figure 4.6: Assembly of electrotaxis device: (a) top view, (b) side view, (c) isometric view

4.3.1 Filling the channels

After the top and bottom plates are assembled, the channels are open only at the ends. This creates a differential pressure between the inside and outside of the channel during filling with culture medium. This difference in pressure leads to the formation of bubbles inside the channel which is not desirable. In order to overcome this problem, the channels are filled by using a standard procedure [11]. To avoid air gaps inside the channels, 3 ml of cell suspension is dropped carefully in one of the openings of the channel. Now, due to the difference in pressure inside the channel, the cell suspension fills one channel completely. The cells are then incubated for 6 h and the liquid is removed using a micropipette attached with a suction pump. A drop (3 ml) of cell culture medium is first used to fill one channel. A 0.3 μl drop is dropped in the same opening of the channel. Now the excess medium fills the second channel filling the entire bottom plate without any air gaps. The dishes are then filled with more medium to ensure the proper growth of cells.

4.4 Cell culture experimental procedures

In order to study the normal behavior of cells, animal cells can be used as a good experimental model. Experiments conducted in vitro on animal cells give a preliminary in vivo understanding of functionality of whole tissues and organs. For cell culturing, cells are first isolated from the required tissue and then seeded into plastic culture flasks and grown in a sterile environment.

Human osteosarcoma cells (SaOS-2), non-transformed cells with osteoblastic properties, were obtained from the American Type Culture Collection (ATCC HTB 85) and cultured in McCoy's 5A medium (Gibco BRL, Karlsruhe, Germany) containing 15% fetal calf serum and 1.25% L-glutamine. 100 ml of 20000 cells/ml was seeded into the channel for EF applications [24].

4.4 Microscopy for monitoring cell migration

4.4.1 Optical microscopy

In order to study the migration behavior of cells, live imaging of cell migration tracks were recorded using an inverted microscope. Photographs were taken every 1 min for 6 h and a video was generated by merging all the photographs. All the biological imagining experiments were carried out using an Olympus IX81 inverted microscope equipped with DIC components and an integrated vital microscopy chamber (Olympus, Hamburg, Germany). For long term cell migration experiments, the environment should be controlled in an optimal range (37 °C and 5 % CO_2). For this reason, the scan area of the microscope is enclosed inside a plexiglass box with temperature and CO_2 level controller. The microscope is connected to a computer where built-in software is used to precisely move the stage for selecting a good area for migration experiments. In order to study the migration of cells, the whole process is automated using the software where it automatically acquires images at fixed intervals and at the end of the experiments all the images are stitched together to produce a video of migration of cells. The software also adjusts the focal plane within the specified range to obtain the best image. The Olympus IX81 microscope is equipped with an inverse optics system, i.e. the light is guided from the top side of the sample and the lens is below the sample. The microscope has three modes of operation:

- Brightfield microscopy: This mode is generally used for thin and transparent samples typically suitable for cell imaging. The contrast is obtained by different light absorption levels in the sample.
- Phase contrast microscopy: In this mode, the difference in contrast is obtained by the different wavelengths emitted by the sample. High magnification imaging is possible in this mode of operation.

- Differential interference microscopy: This is a type of phase contrast microscopy in which the sample is imaged based on the gradients of the wavelengths emitted by the sample.

Additionally, the microscope is equipped with different apertures (10x, 20x and 40x) for different magnifications, ultra violet light source and different color filters for performing fluorescence dye staining experiments.

4.4.1. Fluorescence microscopy

Fluorescence microscopy is a method in which the distributions of fluorescent particles are observed in biological cells. In this method, fluorescent dye which is added to the cell suspension, is used and functionalized to bind to specific targets in a cell [88]. When these particles bind to specific sites, ultraviolet light is used to excite these particles and by using color filters and defined wavelength, the distribution of these particles can be studied. The intensity and wavelength of the light emitted by the particle depends on the binding of the particles to the target [88]. The emitted light is then captured on a CCD chip of the camera. There are two major classes of ion-sensitive dyes:

- Single-wavelength dyes: the intensity of the fluorescence emission increases proportionally with the free-ion concentration. The problem with these dyes is that it is difficult to distinguish between differences in ion concentration and variation in dye brightness caused due to dye concentration and dye photobleaching.
- Dual-emission ratio metric dyes: These dyes can be excited at two different wavelengths. The emitted dye increases in concentration at the excitation (first) wavelength and decreases at second wavelength.

In all the fluorescence experiments performed, ratio metric calcium dye (Fura 2 AM, catalog # F-1221) was used.

4.4.2. Phase contrast microscopy

Bright-field microscopy cannot detect the details in a living cell because of the low contrast between structures with similar transparencies and insufficient natural pigmentation. To overcome this problem, a special form of optical microscopy can be used called the phase contrast microscopy. The principle of operation is based on the effect that high refractive structures bend light greater than structures that have a lower refractive index. It is particularly useful because the organelles have a wide range in refractive indices. A more specialized method in phase contrast microscopy is called differential interference contrast microscopy (DIC). In this method, the light intensity of the image corresponds to the gradient of the optical wavelength of the specimen in focus. In this method, the edges are strongly highlighted resulting in a realistic image. In order to obtain the greatest depth of field, lenses with high numerical aperture (*NA*) should be used. The relationship between numerical aperture and depth of field (*Z*) is given by:

$$Z = \frac{n\lambda}{2NA^2} \tag{4.1}$$

where n is the refractive index of the culture medium and λ is the wavelength of the light source [89]. In this study, a lens with 20x aperture is used (*NA*=0.4) which gives a depth of field of 58 μm. The 20x aperture was selected to observe a minimum of 20 cells in the field of view.

4.5 Cell migration quantification

In this section the steps involved in analyzing the cell migration tracks are outlined. ImageJ (NIH software) is used for analyzing the cell migration tracks. A built-in plug-in in this software allows for manual tracking of the cell. Once the tracks of the cells are manually selected the corresponding coordinates are transferred to Excel software (Microsoft Inc.). The rectangular coordinates are then translated into polar coordinates. They are then plotted in Origin Pro 8 software. The cell migration tracks are analyzed as follows:

Step 1: Importing the image sequence into Image J software. Figure 4.7 is an image taken from the software screen. It shows the image of a video sequence at the start of the experiment.

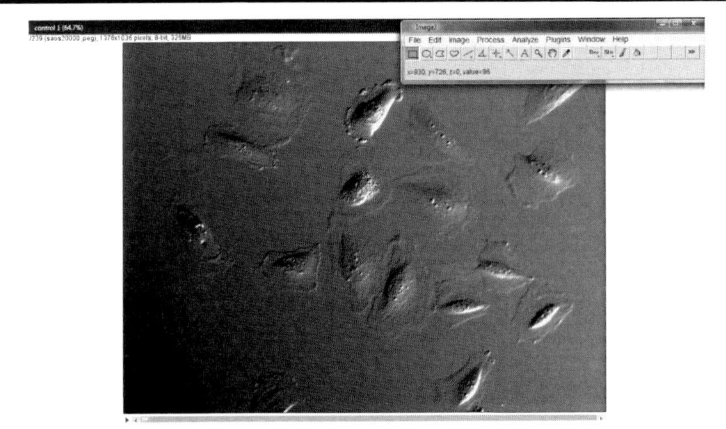

Figure 4.7: Photograph showing the image sequence loaded in the software ImageJ

Step 2: Using the manual tracking plug-in in the ImageJ software to track the cells individually. The path taken by the cell is depicted in Figure 4.8. The beginning of the path (Figure 4.8a) and the end of the path (Figure 4.8b) are shown. The box to the right bottom corner shown in Figure 4.9a, b displays the coordinates of the points traced by the cell.

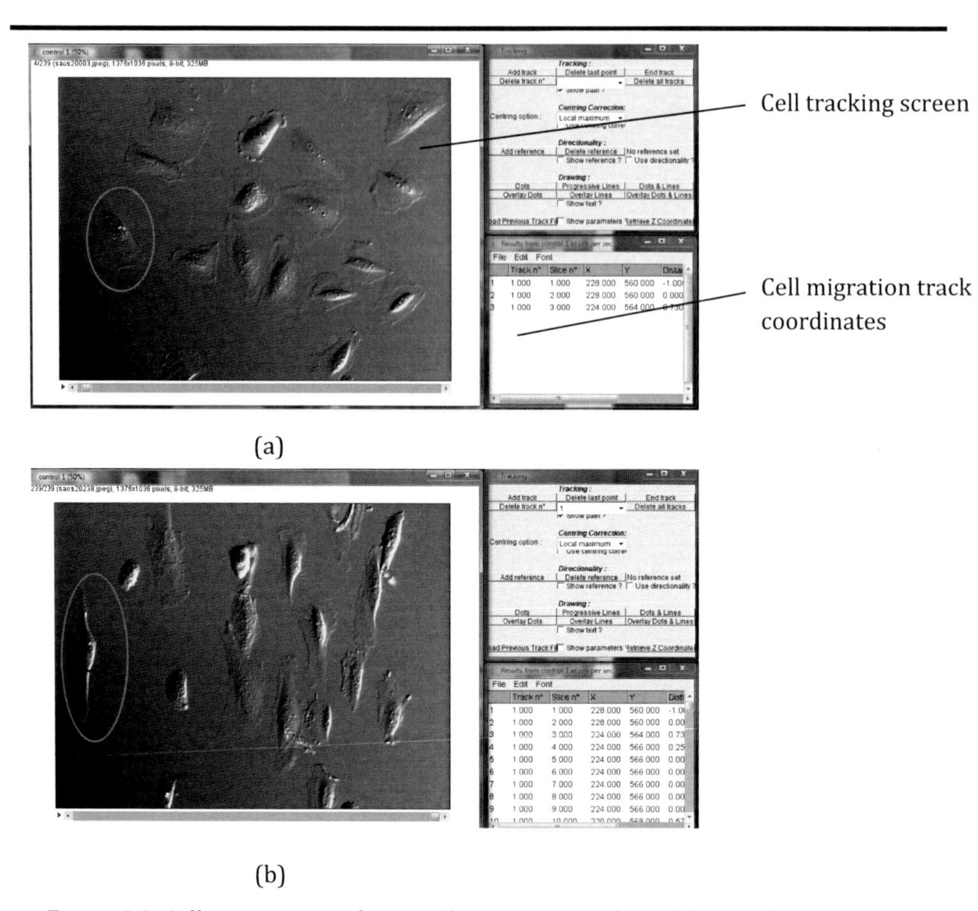

Figure 4.8: Cell migration analysis. Cell migration tracking (a) at the beginning of the and (b) at the end of the experiment.

Step 3: The same procedure explained in step 2 is repeated for all the cells and the coordinates are saved in an Excel file.

Step 4: The coordinates are then transformed to a fixed coordinate system where the center (0, 0) is fixed at the center of the image. This transforms all the tracks of the cells to a standard coordinate system.

Step 5: The rectangular coordinates are then transformed in to polar coordinates for a better visualization of the cell migration tracks. The polar coordinate system can be used to compare the migration behavior of one cell with respect to another.

Step 6: The polar coordinates of all the cell migration tracks are then plotted in Origin Pro 8 software. A polar plot is shown in Figure 4.10. The polar plot gives information on the angle of migration as well as the distance. From this plot, the speed of migration can be computed.

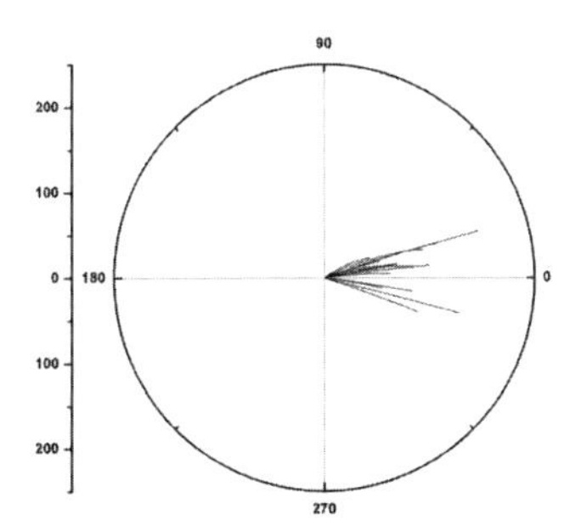

Figure 4.10: Polar plot of cell migration data of several cells.

5 Electric field strength characterization

The cells migrate in response to the *EFS*. In order to understand this behavior it is important to understand the distribution of *EFS* inside the electrotaxis device. In this chapter the variation of *EFS* is studied in both homogenous and non-homogenous electric fields. The dependence of the *EFS* on channel parameters like length, breadth and height are studied. The *EFS* is simulated and these results are verified experimentally.

5.1 Ohmic power dissipation in microfluidic devices

EFS is the force experienced by a unit positive charge at that point. It is a vector quantity and has units of V/mm. Based on the *EFS* values, we have two configurations: homogenous and non-homogenous. In homogenous configuration, the *EFS* is constant with respect to the distance from the electrodes whereas in a non-homogenous configuration, the *EFS* varies with respect to the distance between the electrodes. Cells respond to the *EFS* what can be studied in a micro-fluidic device.

Ideally, for a given voltage in a system, the *EFS* does not depend on the height and width of the channel and is only inversely proportional to the length of the channel. Due to the considerable specific resistivity of the fluid, Joule heating takes place, affecting the relationship between the width and height of the channel to the *EFS*.
The Ohmic power dissipation P_{loss} through the conducting liquid medium is given by:

$$P_{loss} = V_{ch} . I_{ch} = \frac{V_{ch}^2}{R_{ch}} \cong G.\Delta T \qquad (5.1)$$

where I_{ch} is the electric current, G the thermal conductivity of the fluid and ΔT the temperature increase due to heating. R_{ch} is the electrical resistance of the channel. From Ohm's law, R_{ch} can be expressed as:

$$\frac{V_{ch}}{I_{ch}} \cong R_{ch} = \frac{\rho_{ch} l_{ch}}{A_{ch}} \qquad (5.2)$$

where ρ_{ch} is the specific electrical resistance of the culture medium, l_{ch} the length of the channel, and $A_{ch} = w_{ch}$ x h_{ch} the cross-sectional area of the channel. Eqs. (3) and (4) yield

$$P_{loss} = \frac{V_{ch}^2 . w_{ch} . h_{ch}}{\rho_{ch} . l_{ch}}. \qquad (5.3)$$

The Ohmic power dissipation P_{loss} inside the channel is directly proportional to the width and height of the channel and to the square of the voltage across the channel length. As the height of the channel is increased, higher current values occur leading to increased heating of the conducting medium because of the higher resistance. So the height of the channel plays an important role in designing an electrotaxis chamber.

5.2 Experimental methods

In order to study the effect of channel dimensions on *EFS* for a homogenous electric field configuration, devices were fabricated with different dimensions in length (l_{ch}), width (w_{ch}) and height (h_{ch}) of channels. The basic design of a micro-fluidic device is shown in Fig. 5.1. It consists of a micro-fluidic channel and a reservoir which is filled with the conducting medium. Polydimethylsiloxane (PDMS) and glass slides were used in this work to fabricate such devices. PDMS (Sylgard 184 Silicone Elastomer, Dow Corning, MI) was mixed in a

10:1 ratio with the curing agent. 3M adhesive sheets (3M 467 MP, 200 M adhesive, 0.6 mm in thickness) were cut out in the required dimensions and used as a negative pattern for fabricating the micro-fluidic channels in these devices. Fig. 5.2 shows the fabrication steps of one of such a device. In the first step, the channel with required dimension is cut and placed at the center of the PDMS mould (Fig 5.2a). Subsequently, the PDMS is poured into the mould with the medium wells placed at both ends of the channel (Fig. 5.2b) and is cured at 50°C for 1 hr (Fig. 5.2c). In the last step, the cured PDMS device is carefully peeled off and a glass slide is plasma-bonded to seal the channel (Fig. 5.2d).

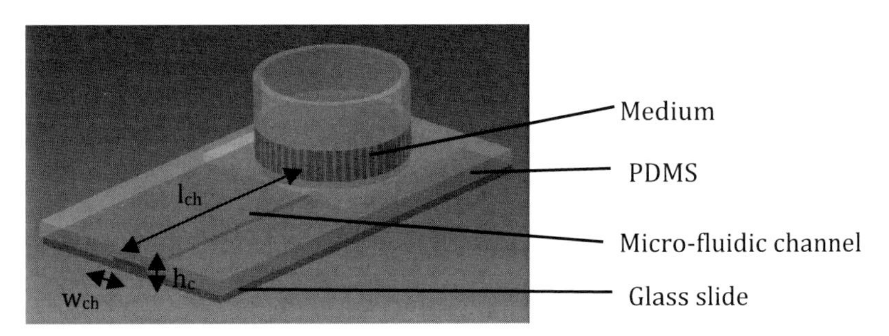

Figure 5.1: Sectional view of a micro-fluidic chip to measure electrotaxis of cells

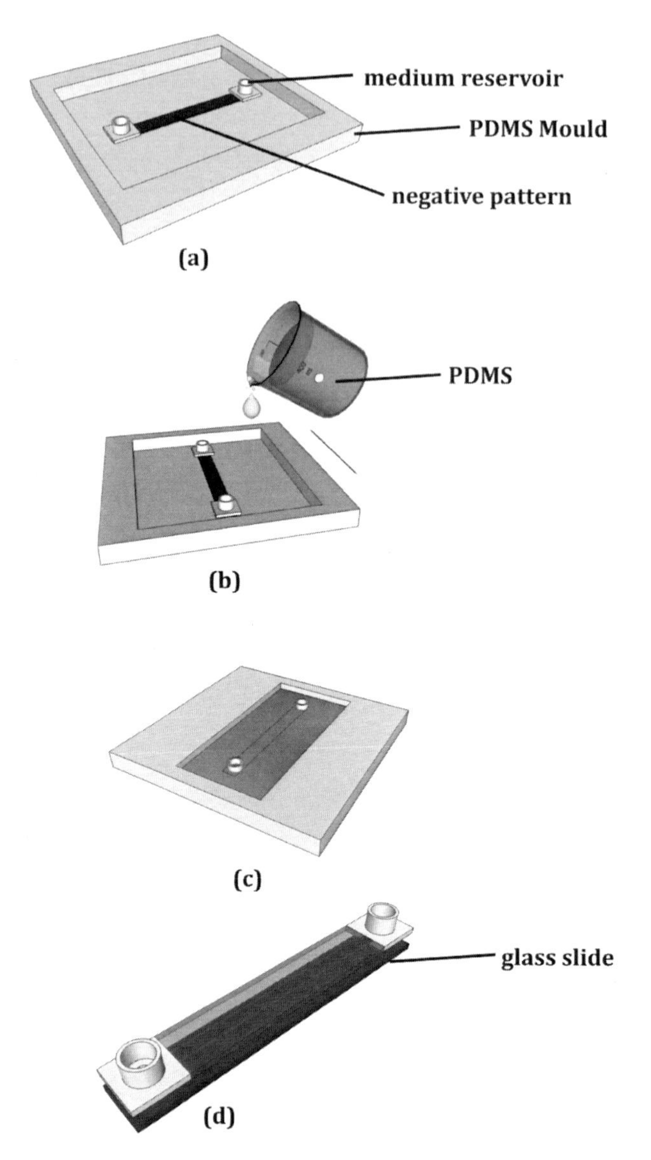

Figure 5.2: Fabrication steps of the micro-fluidic devices to measure *EFS* experimentally (explanation in text)

The influence of all the three parameters, l_{ch}, w_{ch} and h_{ch}, on *EFS* was studied. For each case, the other parameters were kept constant (Table 5.1).

Parameter studied	Fixed parameters
l = 25 mm, 50 mm, 100 mm, 50 mm (S-curve)	w = 5 mm, h = 0.6 mm
w = 2.5 mm, 5 mm, 10 mm, 15 mm	l = 50 mm, h = 0.6 mm
h = 0.6 mm, 1 mm, 2 mm, 4 mm	l = 50 mm, w= 5 mm

Table 5.1: Experimental plan for studying the relationship between *EFS* and the channel dimensions

The *EFS* was simulated and verified experimentally. Electronet (Infoanalytica Inc.) simulation software was used to simulate the *EFS* inside the channels. In this work, homogenous *EFS* was not directly measured. Two leads of a multimeter (Voltcraft VC160) were inserted into two different medium reservoirs to measure the voltage V_{ch} over the channel and this value of voltage V_{ch} was divided by the distance between the leads (which is the length of the channel, l_{ch}) to obtain the *EFS* as shown in Eq. (5. 2). The supply voltage V_{supply} was adjusted such that V_{ch} amounted to the targeted voltage. It should be noted that, in order for the cells to grow and multiply for a long term experiment (more than 24 hours), a minimum height of 0.4 mm of the channel is required. Based on this data the experiments were designed.

Two configurations (Figure 5.3) were tested for non-homogenous *EFS*. In Fig. 5.3a the positive and negative electrodes are alternated. In Fig. 5.3b the positive and negative electrodes are placed consecutive by the *EFS* were first simulated and then verified experimentally.

Figure 5.4 shows the experimental setup for measuring *EFS* experimentally. The setup consists of a 52 pin array chip with the leads placed 2 mm apart. This chip is inserted into the device and sealed from the top. Each pin is connected to a bread board via wires soldered to them and the output is read out in a computer using a National Instruments Data Acquisition Unit (NI USB 6229). Labview software is used to measure the voltage difference between two pins. This voltage is divided by the distance between corresponding pins to obtain the value of the *EFS*.

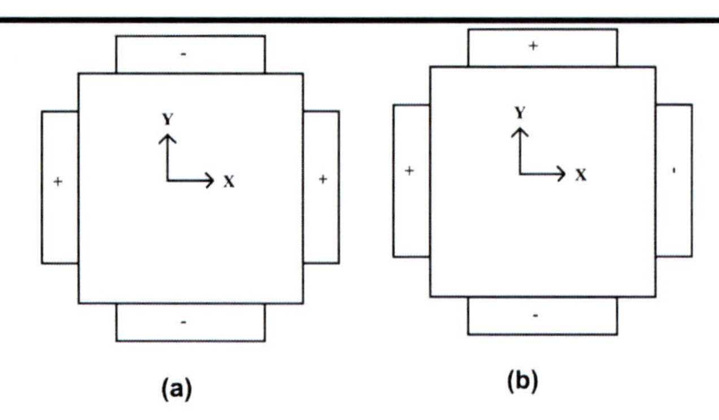

Figure 5.3: Non-homogenous electric field strength configuration: (a) alternating electrodes, (b) consecutive electrodes

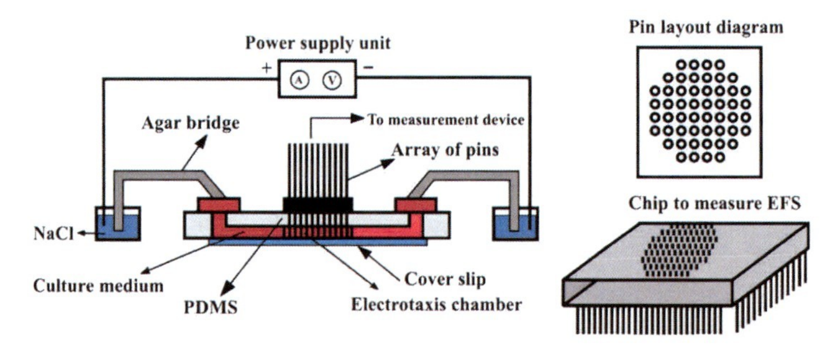

Figure 5.4: Setup to measure non-homogenous EFSs

5.3 Homogenous *EFS*

5.3.1 Simulation of the *EFS*

Simulation was run in the simulation software using a basic model (Fig. 5.5) consisting of two electrodes and the channel of the required dimension. Both electrodes contact the complete cross-section of the micro-fluidic channel. The thermal conductivity of PDMS material was assumed as 0.15 W/mK [Polymer Data handbook 2009]. The relative permittivity and conductivity of the cell culture medium, (Dulbeccos Phosphate Buffered Saline (D-PBS)) were fixed as 80 and 1.6 Ω/m, respectively, as described in [Lee W. D. and Cho Y.-H. 2007]. The whole model is enclosed inside a PDMS material whose permittivity and conductivity are negligible compared to the cell culture medium. The simulation is performed using a static 2d model available in the software.

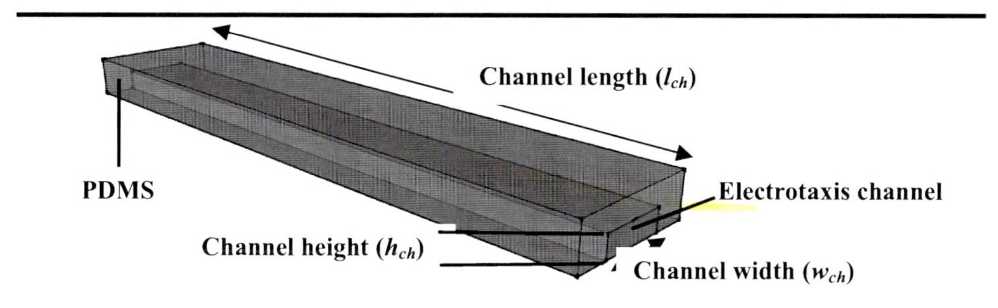

Figure 5.5: Basic model used for simulating the *EFS*

Figure 5.6 shows the simulation results of different lengths of the channel. Results show a linear relationship between voltage V_{ch} and length l_{ch} if *EFS* is maintained constant. An inversely proportional relationship between *EFS* and l_{ch} occurs, if V_{ch} is maintained constant as expected from Eq. (5. 2).

Electrotaxis has been investigated in curved micro-fluidic channels [90]. To study the effect of curvature on *EFS* a "S-curve" was investigated. Fig. 5.6d shows the simulation results of a channel in the shape of an "S". The length (center line of the S-curve) and width of the S-curve was fixed as 50 mm and 5 mm, respectively. We found that the electric field lines are slightly deformed inside the channel. The small deformations could be due to the numerical algorithm used in simulation. From the data produced by the simulation, voltage was plotted along two different locations of the S-curve (Fig. 5.7). Due to symmetry of the S-curve, only half the length of the S-curve was studied. Voltage variations along the edge of the S-curve and at the center of the S-curve were considered. The voltage varied linearly with respect to the length of the channel for the curve along the edge of the channel. The voltage had minor fluctuations along the center of the S-channel. These fluctuations were due to approximation of curve lengths. The average slope of the lines in Fig. 5.7 were found as 0.411 V/mm and 0.288 V/mm for the curve along the edge and the curve at the center of the channel, respectively. So the *EFS* is lower inside the channel with respect to the edges.

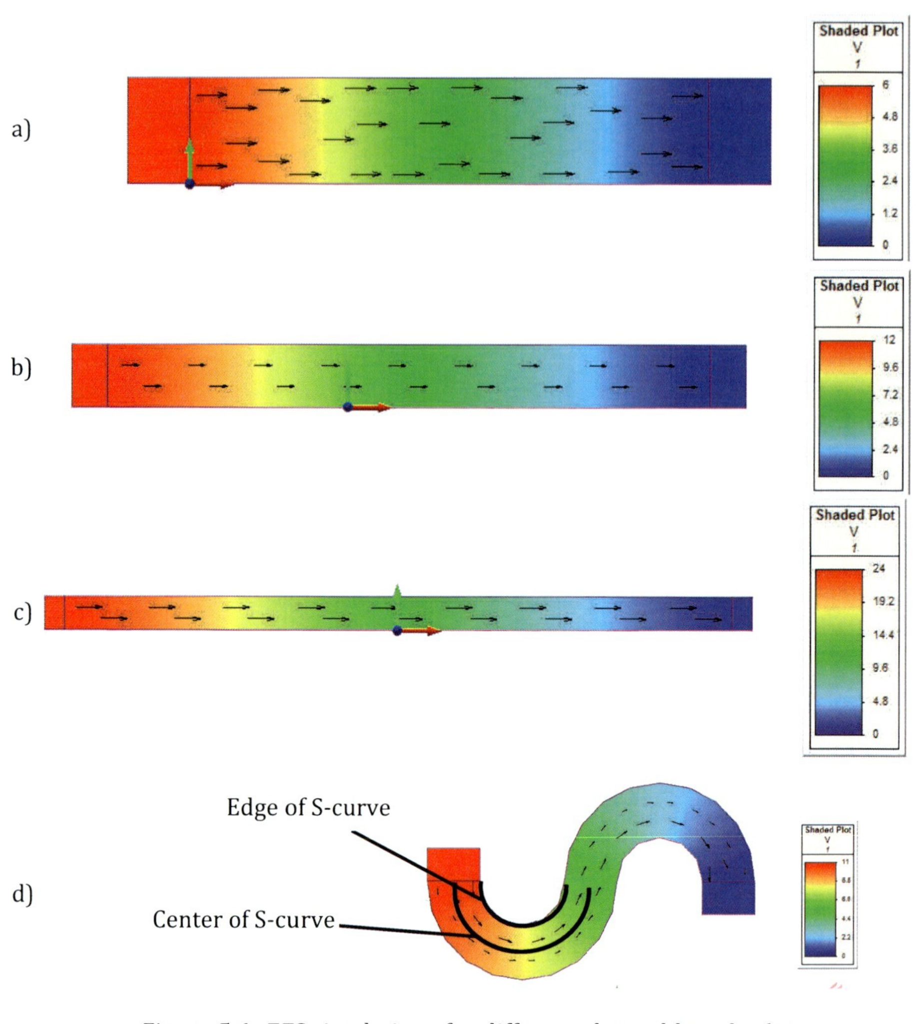

Figure 5.6: *EFS* simulations for different channel lengths (see table 1)

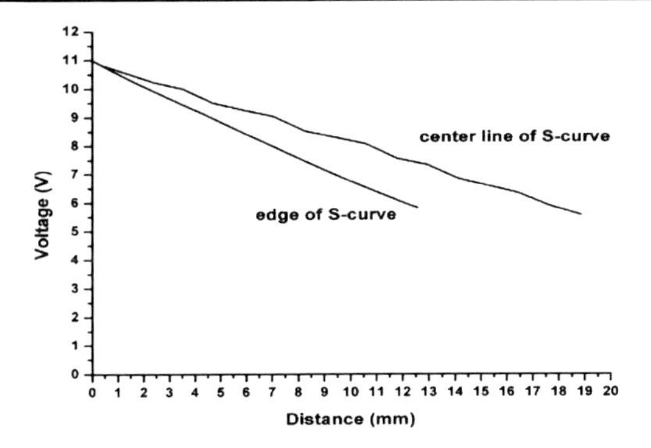

Figure 5.7: Variation of voltage at different points for a S-shaped curve

The effect of width of the channel on *EFS* is shown in Fig. 5.8. *EFS* was inversely proportional to the width of the channel. The *EFS* is lower for channel widths larger than the electrode compared to the channel widths equal to and less than the electrode dimension. The electric field lines in the case of larger channel widths are slightly bent near the electrodes. The *EFS* is dispersed over a larger width which results in a lower value.

5.3.2 Experimental determination of *EFS* in micro-fluidic channels

In order to verify the simulation results, *EFS* was experimentally measured as described in section 5.2. Fig. 5.9 shows data plotted for simulated and experimental values of *EFS* for different length (Fig 5.9a) and width (Fig. 5.9b) of channels. The experimental and simulated data match very well. The measurement uncertainty was less than 2 %. In Fig. 5.9c, *EFS* is plotted against height of the channel for two different voltages (50 V and 100 V) supplied at the channel ends. The *EFS* decreased exponentially with the increase in height of the channel.

This could be because of the Joule heating effect inside the channel. The heat Q is transferred from the culture medium to the PDMS:

$$P = \frac{V_{ch}^2}{R_{ch}} \tag{5.4}$$

$$Q = \frac{V_{ch}^2 \cdot t}{R_{ch}} \tag{5.5}$$

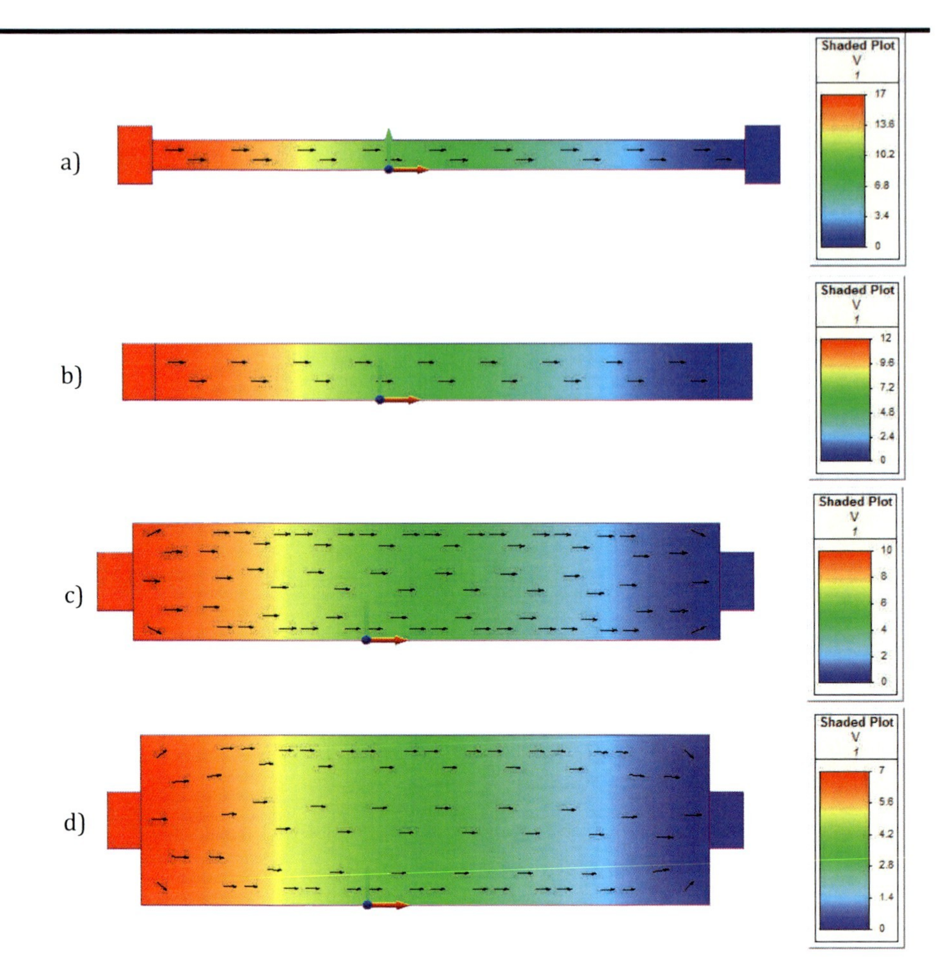

Figure 5.8: Voltage (V_{ch}) vs width (w_{ch}) of the channel (see table 5. 1)
a) w = 2.5 mm, b) w = 5 mm, c) w = 10 mm, d) w = 15 mm

$$Q = m.s.\Delta T \tag{5.6}$$

where P is the power generated in a circuit, Q the heat generated, t the time, m the mass of conducting fluid, s the specific heat conductivity of medium and ΔT the rise in temperature.

Using equations (5.4), (5.5) and (5.6), it yields

$$\Delta T = \frac{V_{ch}^2.t}{R.m.s} = \frac{P.t}{m.s}. \tag{5.7}$$

When the height of the channel (h_{ch}) increases, higher voltage (V_{ch}) is required to maintain constant *EFS*. So from equation (5. 9) we can observe that an increase in V_{ch} increases the temperature of the system. This explains the rise in temperature due to the increase in channel height.

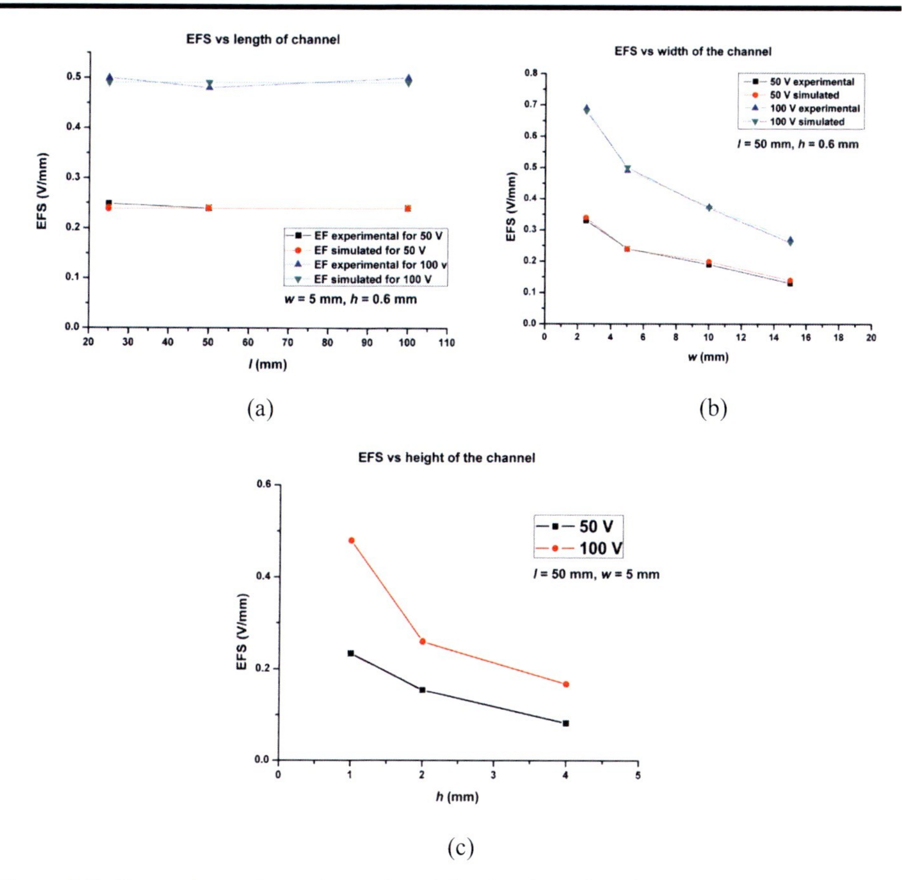

Figure 5.9: Comparison of experimental and simulated results of *EFS* vs geometric dimensions

Figure 5.10 shows the change in temperature inside the micro-fluidic channel for different heights of the channel. A constant *EFS* (measured) of 0.5 V/mm was applied in all the cases by controlling V_{ch}. The temperature *T* was measured for 5 h using an infrared thermometer (Raynger MX, Raytek Inc.) for every 10 min inside the micro-fluidic channel. The temperature requirement for an electrotaxis experiment is very crucial. The living cells are studied at the nominal temperature of (37 ± 1) °C. Any temperature higher than this would change the cell characteristics decisively. From Figure 5.10 it is clear that for heights above 1 mm, Joule heating generates a considerable change in temperature unsuitable for electrotaxis experiments. The thermal conductivity of the culture medium was found to be 0.58 W/mK [91]. From Eq. (5), when the height of the channel increases, the power loss P_{loss} increases which increases the thermal conductivity of the culture medium. So there is a raise in temperature for higher channel heights when all other parameters are maintained constant.

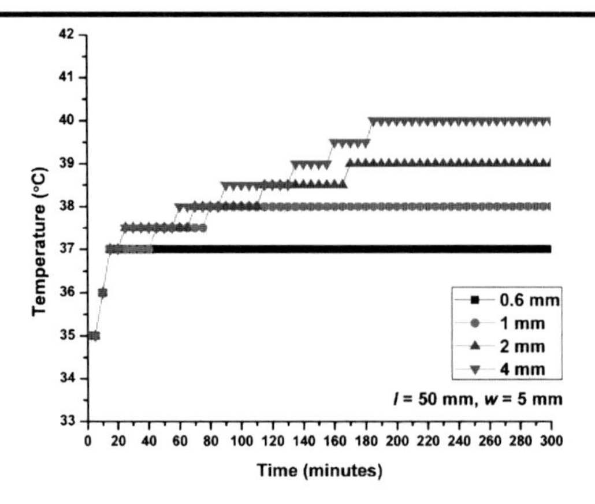

Figure 5.10: Joule heating effect for different heights of the

5.4 Non-homogenous *EFSs*

5.4.1 Simulation of non-homogenous *EFSs*

The *EFS* varies with respect to the distance from the electrode in a non-homogenous configuration. This is of particular interest because the cells would experience a gradient of *EFS* inside the chamber. The *EFSs* were first simulated (Electronet software) and then verified experimentally. Figure 5.11 shows the basic model used for simulation. It consists of a cell chamber (15 mm x 15 mm) and four electrodes marked as E1-E4 clockwise. Two configurations were tested using this model. In the first configuration, the positive and negative electrodes are alternated. In the second configuration, the positive and negative electrodes are consecutive.

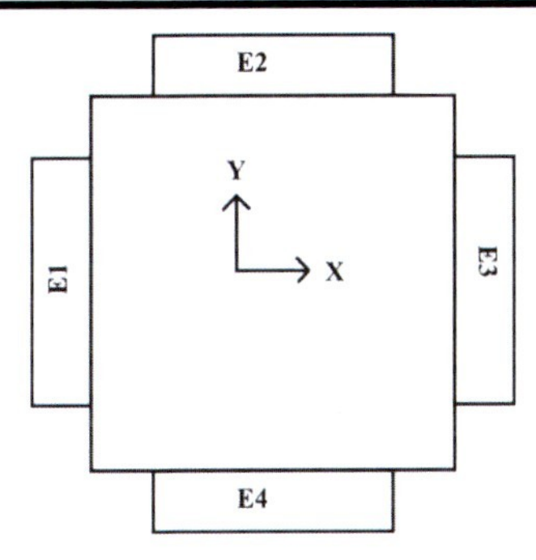

Figure 5.11: Basic model used for simulating non-homogenous electric field strength

Figure 5.12 shows the simulation results of the two configurations investigated for non-homogenous *EFSs*. The simulation was carried out for an applied voltage of 50 V. In the first configuration of figure 5.12a, the positive and negative electrodes are alternated. The electric field lines were observed to bend starting from the positive electrode towards both the negative electrodes. This would create larger *EFS* at the corners and lower at the center of the chamber. The *EFSs* are symmetric with respect to one quarter of the chamber.

In configuration of figure 5.12b, the positive and negative electrodes are consecutive. The simulations were carried out for a voltage of 50 V. The field lines are directed towards the negative electrodes. The *EFS* is not symmetric in this setup. It can be observed that the *EFS* is larger near the positive electrodes and decreases near the negative electrodes.

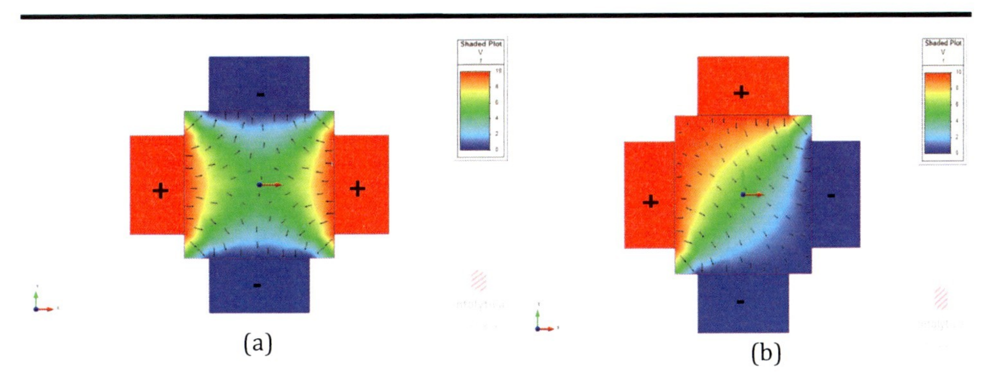

Figure 5.12: Simulation result for (a) configuration 1 and (b) configuration 2 with non-homogenous EFSs

5.4.2 Experimental verification of non-homogenous *EFSs*

In order to confirm the simulation results obtained for non-homogenous *EFSs*, the *EFS* inside the chamber was measured experimentally. The experimental set-up for measuring the *EFS* is explained in section 5.2. Two non-homogeneous electric field configurations were tested as shown in Figure 5.3. Figure 5.13 shows the experimentally measured values for the two non-homogenous electric field configuration. The electric field lines are plotted in Figure 5.13a and c follow the same pattern as observed in the simulation result. The magnitude of *EFS* is plotted in Figure 5.13b, d. For configuration# 1 it can be observed that the *EFS* is higher at the positive electrode ends and decreases towards the center. The lowest *EFS* was found to be near the negative electrodes. This *EFS* map agrees well with the simulated values. The reduction in *EFS* inside the chamber may be due to the resistance of the culture medium.

In figure 5.13b, the electric field lines start from the positive electrode and move towards the negative electrodes. This result is in accordance with the simulation results shown in figure 5.13d. The magnitude of *EFS* for this configuration is plotted in figure 5.13d. The positive electrodes in this figure were fixed at the back end of the graph (shown in figure 5.13d). The magnitude of *EFS* is higher towards the positive electrodes and decreases as we move towards the negative electrodes. The decrease in *EFS* is due to the resistance in the cell culture medium. This resistance lowers the flow of current inside the chamber.

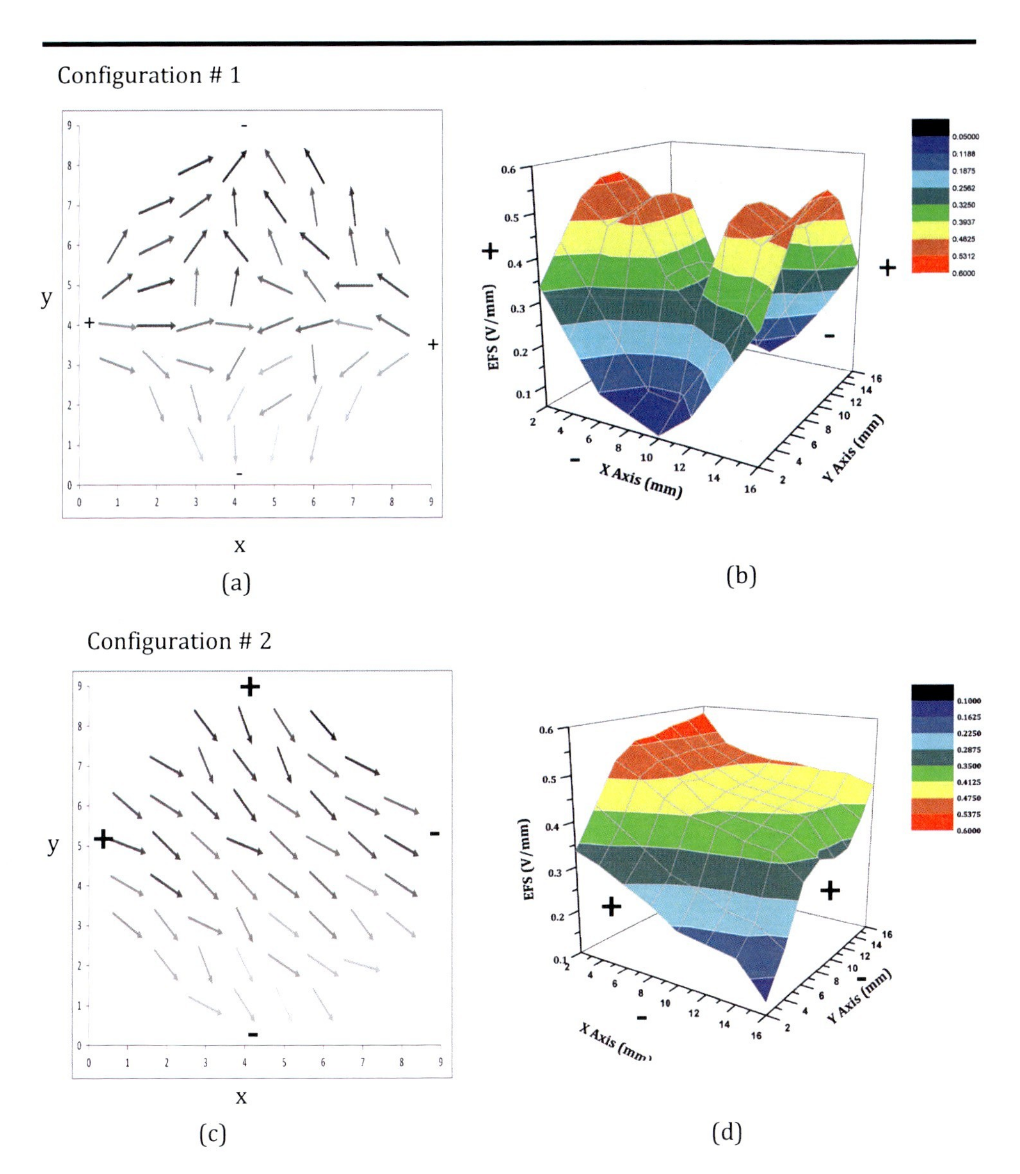

Figure 5.13: Experimental results of non-homogenous configuration #1 and #2 (a, c) direction of electric field lines, (b, d) magnitude of electric field strength

5.5 Significance of simulation data

It is very important to verify the accuracy of the simulation results for two reasons:

- it helps in understanding the electric field gradients and directions of electric field lines for values not measurable experimentally.
- the observable area inside the cell culture chamber is less than 1 mm. For such areas it is not practical to determine the *EFS*s and directions of electric field lines experimentally. The simulation values can thus be used for such small areas.

The simulation and experimental values of *EFS* magnitudes are plotted in figure 5.14 to determine the accuracy of the simulated result for configuration 1 and configuration 2 respectively. For configuration 1, the *EFS* values along the horizontal line at the center of the chamber were computed and plotted. In configuration 2, the *EFS* were computed and plotted along the diagonal length of the chamber. The simulated and experimental data match closely for both the configurations. The mean deviation was calculated as 2 %. This deviation is due to the limited number of points to reconstruct the *EFS* map accurately. Nevertheless, the margin of error is acceptable to closely predict the *EFS* and the direction of field lines. This result confirms the accuracy of the simulated values and can thus be used for quantifying the cell migration direction and speed.

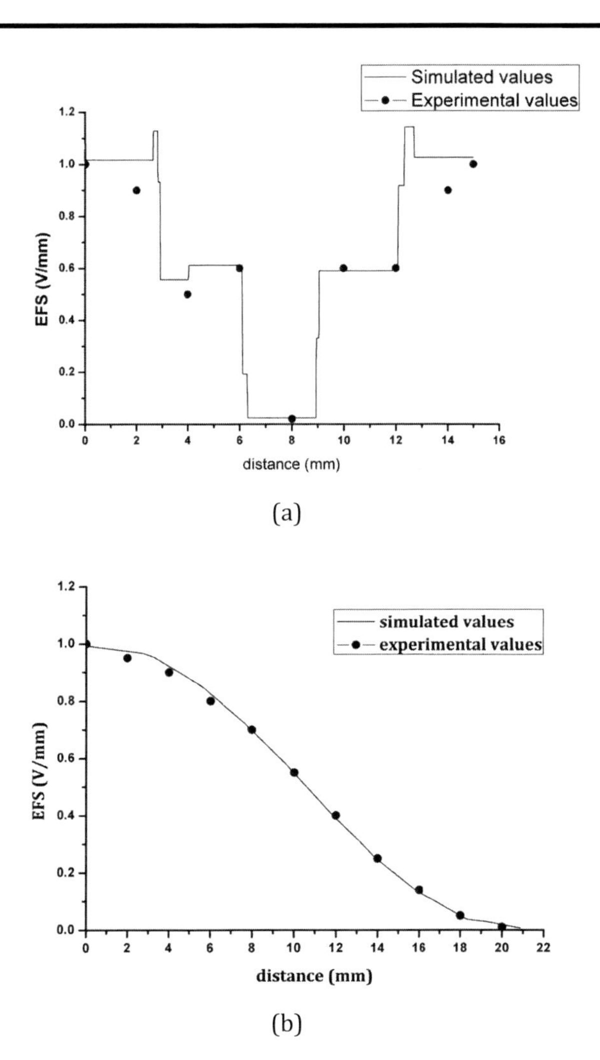

(a)

(b)

Figure 5.14: Verification of simulation data for configuration (a)#1 and (b)#2 from figure 5.13

6 Results and Discussions

In this chapter, the biocompatibility of the device is first tested. Results on cell spreading and attachment are presented. The experimental procedure and how the results are quantified are outlined. Results on the control, homogenous and non-homogenous electric field configurations are presented. In the control experiment, no *EFS* is applied and the cells showed random migration behavior, whereas in homogenous and non-homogenous *EFS* the cells showed directed migration with enhanced migration speed.

6.1 Biocompatibility results

In order to evaluate the biocompatibility of the device, Human osteosarcoma cells (SaOS-2), were then incubated for 24 hours in PDMS channels that were formed on thin cover slips. Immunoflourescence labeling was performed on the cells to visualize the cell attachment, cell spreading, cytoskeleton organization and focal adhesion formulation. To examine the focal adhesion, cells were incubated with a mouse monoclonal antibody against human vinculin (dilution 1/100, Serotec, UK) overnight at 4 °C (shown as green in Fig. 6.1a) and FITC coupled goat anti-mouse antibody (dilution 1/100, Dianova, Germany) for 1 h at room temperature. To observe the filamentous actin cytoskeleton, cells were additionally incubated with TRITC conjugated phalloidin (dilution 1/100, Sigma, Germany) for 1 h at room temperature (shown as red in Fig. 6.1b). Finally, cells nuclei were visualized by incubating the cells with DAPI (dilution 1/20, Sigma, Germany) for 1 h at room temperature (shown as blue in Fig. 6.1c). The cells cultured in the cover slips showed good attachment, spreading and elongated distribution of actin filaments with good focal adhesion formations.

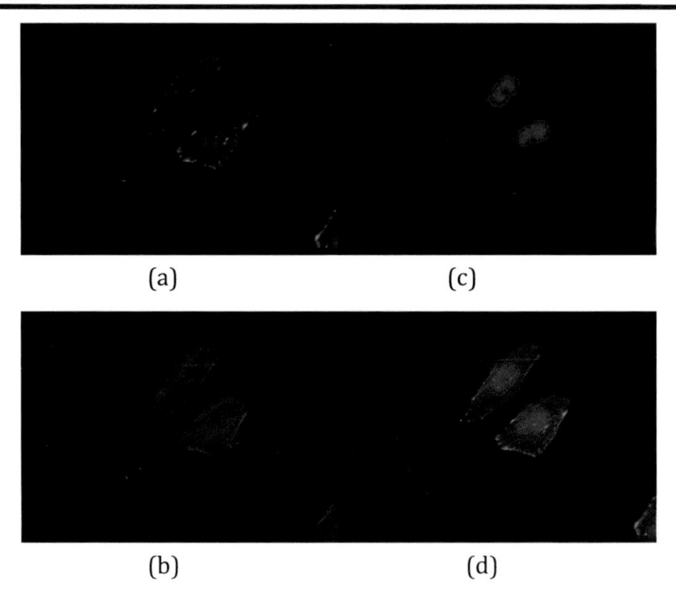

Figure 6.1: Biocompatibility result of the electrotaxis device: (a) focal adhesion using vinculin (b) actin cytoskeleton using phalloidin (c) cell nucleus by DAPI (d) merged images

6.2 Experimental procedure for electrotaxis of cells

This section explains the detailed experimental procedure followed to carry out the cell migration experiments. A flowchart depicting the process is shown in figure 6.2. The procedure involves seven steps.

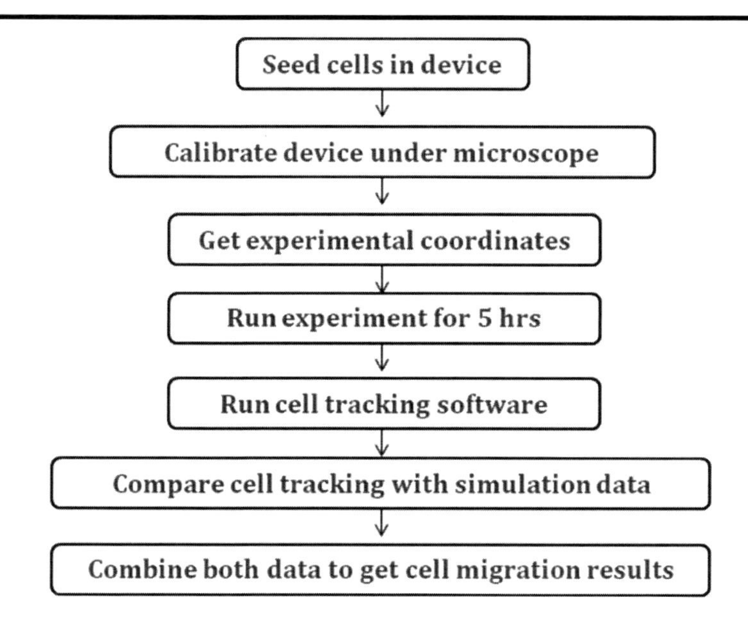

Figure 6.2: Flowchart showing the steps involved in carrying out an electrotaxis experiments

Step 1: Cell seeding

The cells are seeded directly into the device with the required concentration. The device is then placed in the incubator for 7 8 h allowing the cells to attach and grow inside the cell culture chamber of the device.

Step 2: Calibration of the device

The device is placed on a movable stage under the microscope. This stage can be controlled using the built-in software (excellence, Olympus Inc.) for the microscope. Before the experiment is started, the cell culture chamber is calibrated. Excellence software is used for obtaining the end point coordinates of the cell culture chamber area. Using these coordinates, the cell culture chamber dimensions can be computed. These coordinates are used as reference points to determine the relative coordinates inside the cell culture chamber area.

Step 3: Determining experimental coordinates

Once the chamber dimensions are known, healthy and populated cell area is searched inside the cell culture chamber. Once such an area is located, excellence software is used to determine the coordinates of this point. These coordinates can then be used to locate the exact area of investigation inside the cell chamber. These coordinates are compared to the simulation data coordinates to determine the *EFS*s at this location.

Step 4 to step 7 is explained in chapter 4, section 4.5.

6.3 The control experiment

In the control experiment, the cells were studied without any application of an *EFS* for 5 h under the microscope. Figure 6.3 shows the photographs taken at the beginning and end of the experiment. A total of 15 cells were analyzed for the migration behavior. In order to make a conclusive remark about the cell migration behavior, this experiment was performed 5 times and an average of 15 cells were analyzed in each experiment. The tracks of the cells are plotted in the polar graph in figure 6.4. In all the experiments carried out the cells showed random migration behavior. Figure 6.4 shows the direction and distance travelled by each cell. With *EFS*=0, the cells migrated randomly. The average speed of the cells was calculated as 10 μm/h. There was also no significant change in shape and structure of the cells.

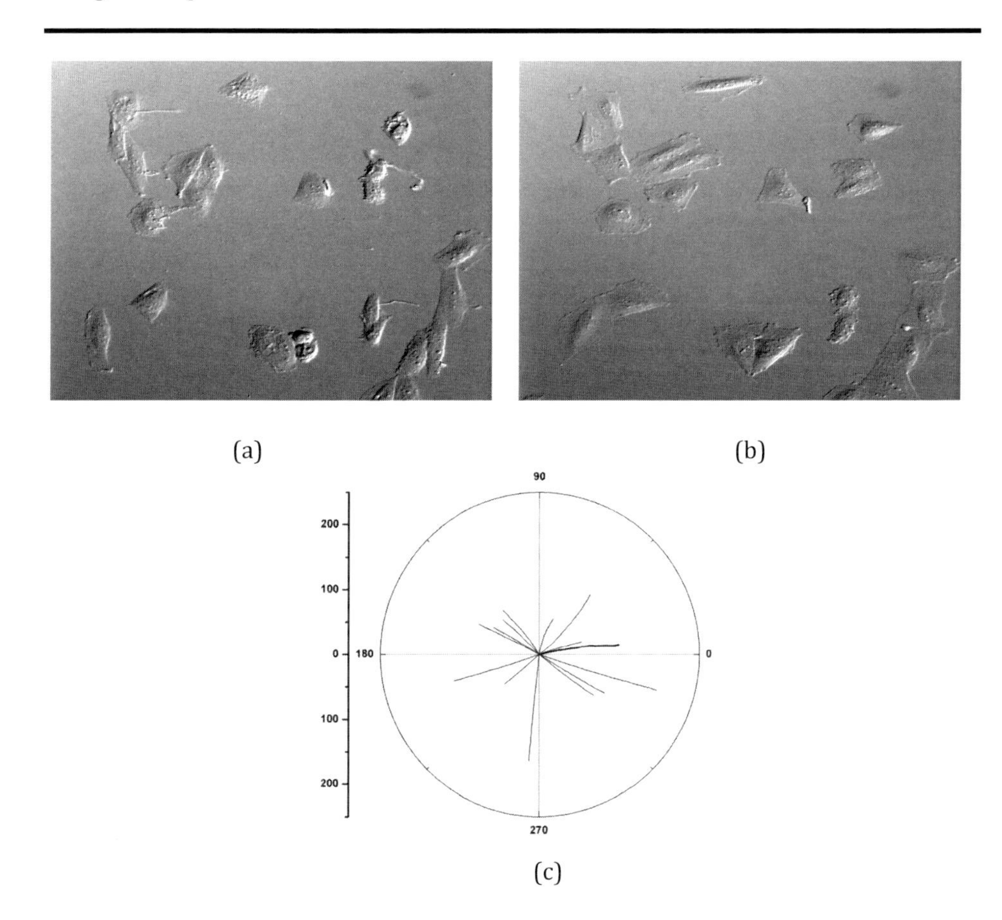

Figure 6.3: Control experiment with *EFS*=0 (a) at t = 0 (b) at t = 5 h (c) cell migration tracks polar plot showing random migration behavior

6.4 Cells under homogenous *EFS*

The cells were exposed to homogenous *EFS*s for 5 h and the migration behavior was captured by the microscope. *EFS*=0.5 V/mm was used to study the migration behavior as this was the reported physiological strength of Saos-2 cells [24]. Figure 6.5 shows the photographs taken at the beginning and end of the 5 h experiment. The cells responded to the *EFS* by elongating perpendicular to the electric field lines (figure 6.6). The cells also showed directed migration towards the anode. It can be seen that all the cells have directed migration. This experiment was repeated 5 times and an average of 15 cells was studied in each experiment. The behavior of cells was consistent in all the experiments. The speed of the cells in homogenous *EFS* of 0.5 V/mm was calculated as 25 μm/h. There was an increase of 400 % in speed with application of *EFS*. This behavior is well reported in the literature as well [23].

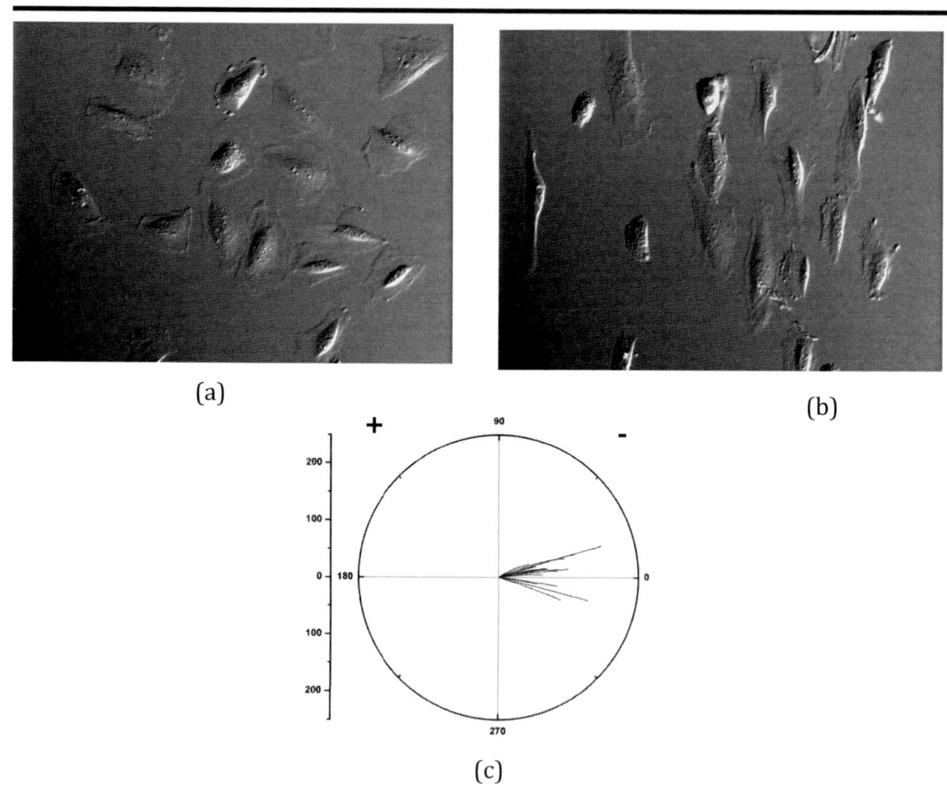

Figure 6.4: Homogenous EFS experiment (a) at time (t) = 0 (b) at t = 5 hours (c) cell migration tracks polar plot showing directional migration behavior

Figure 6.7 shows the variation of migration speed of cells at different *EFS*s [7]. The migration speed increases as the *EFS* is increased. The migration speed reaches a maximum at 1.4 V/mm. The cell experiences high *EFS* beyond 1.4 V/mm and the cellular functions are disrupted. Most of the cells were dead above 1.4 V/mm.

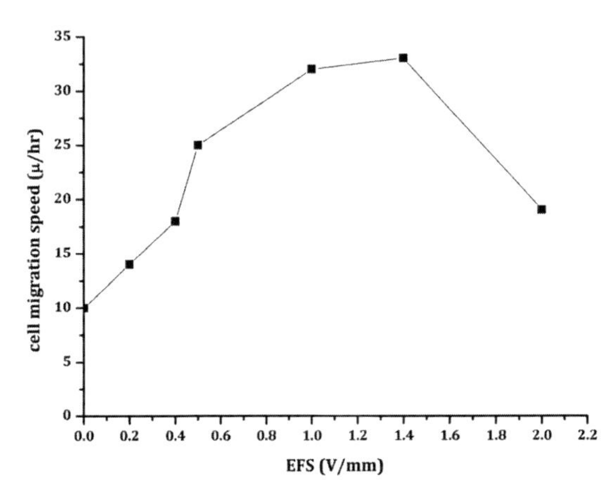

Figure 6.7: Dependence of cell migration speed and *EFS*

6.4 Cells under non-homogenous *EFSs*

Cells were seeded in the device and the non-homogenous configuration #1 was tested. In this configuration, the positive and negative electrodes are alternated (figure 5.12). Before the experiment was started, the device was calibrated under the microscope. The coordinates of the cell culture chamber were obtained from the built-in software of the microscope. Then the coordinates of the point of observation inside the chamber were obtained from the built-in software. The experimental results of the cell experiments with the non-homogenous electric field configuration #1 are shown in figure 6.9. The cells show directed migration within an angular range of (30 42) °. The speed of migration calculated from figure 6.9c was uniform but much slower (16 μm/h) compared to homogenous *EFS*. In order to quantify these results it is important to know the *EFS* and direction of electric field lines experienced by the cells.

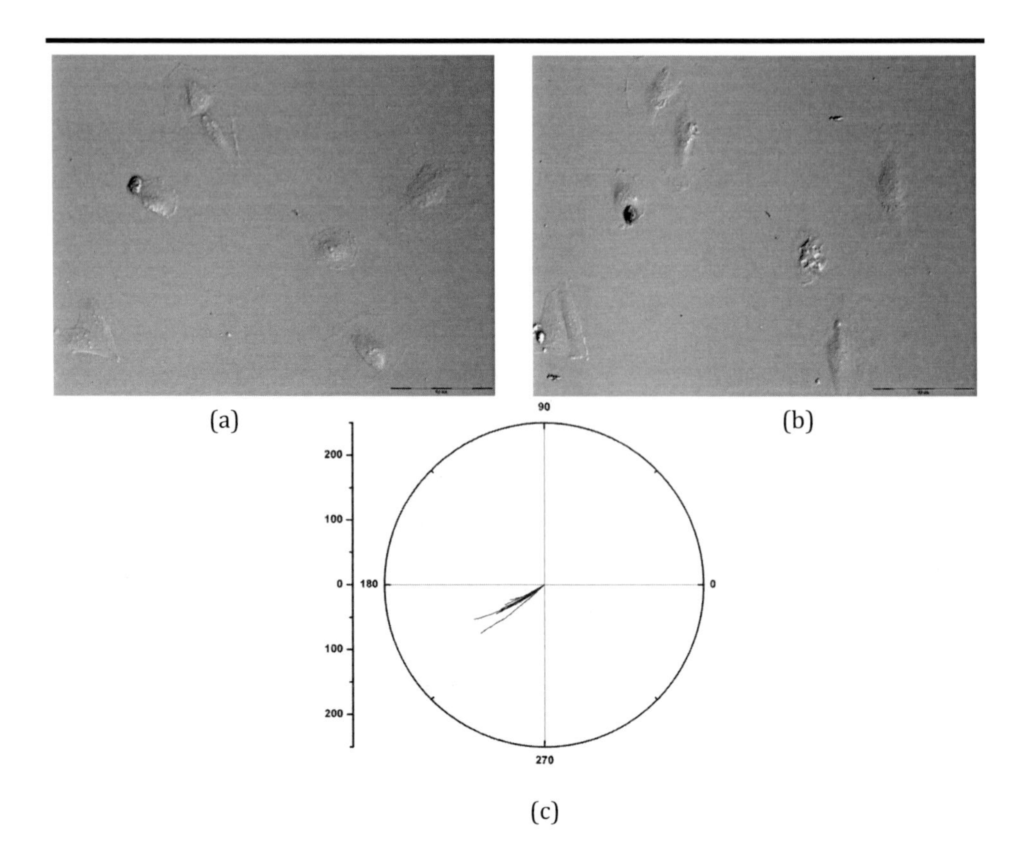

Figure 6.9: Experiments with cells under non-homogenous *EFS* configuration #1 at 0.5 V/mm (a) at t = 0 (b) at t = 5 h (c) cell migration tracks polar plot showing directed migration behavior

Figure 6.10a shows the position of cells inside the cell chamber on a simulation map. Since a 20x objective was used to observe these cells the area of observation is about 1 mm^2. From this figure the angle of the electric field lines varied between (25 35) °. Figure 6.10b shows the magnified view of the simulated *EFS* map for 1 mm^2. The variation of *EFS* is shown in figure 6.11. The *EFS* varied between 0.8...1.1 V/mm. The cells are not able to distinguish *EFS* for such small variations but migrated in the resultant direction of the *EFS*. This result is particularly important in controlling the migration direction of cells just by manipulating the direction of *EFS*s.

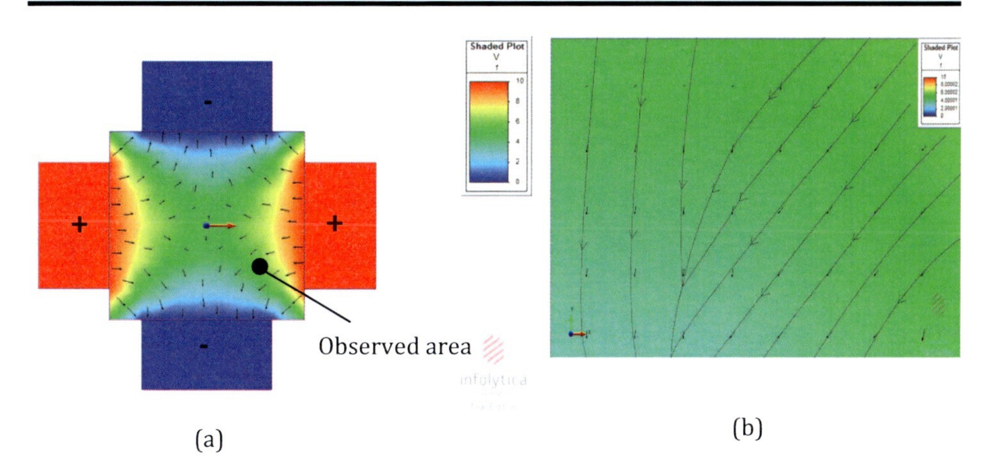

Figure 6.10: (a) observed area inside the cell culture chamber projected on *EFS* map (b) zoomed view of *EFS* map

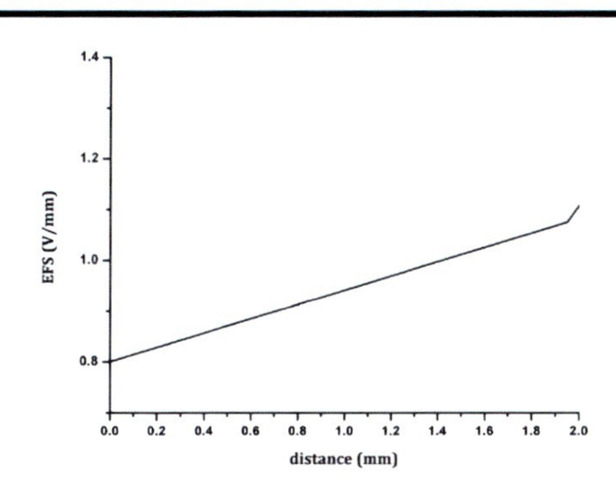

Figure 6.11: variation of *EFS* in the observed area of the experiment

7 Conclusions and future direction

7.1 Conclusions

Cell motility is an important property in all living organisms. Many crucial biological processes involve this process. Directional cell migration may be influenced by various cues which include chemotaxis, hapotaxis, Durotaxis and electrotaxis.

Electrotaxis is the directed migration of cells under an influence of direct current electric field. This phenomenon is essential for many biological processes like embryonic growth, nerve regeneration and wound healing. At least 14 cell types, amoeba and slime worms have been observed to show this property. Although there has been considerable research in this field over the past 100 years we know very little about this behavior in cells. All of the past work focused on studying these cells in electrotaxis chambers, which generated homogenous electric field strength. The cell experiences constant electric field strength in homogenous electric field strength configuration. In this work, a novel designed electrotaxis chamber is explained which is capable of producing non-homogenous electric field strength. The device showed good biocompatibility and electric field strength simulations inside the device matched closely with the experimental values. The main conclusions of this work are summarized as follows:

- Electrotaxis is the directed migration of cells under an influence of direct current electric field strength towards an electrode.
- Electric field strength is constant in homogeneous *EFSS* configuration whereas it varies with respect to the distance from the electrode in non-homogenous *EFSS* configuration.
- Direct current electric field produces products near the electrodes, which are toxic to the cells. Agar salt bridges are used to slow down the transfer of these products to the cells.
- A novel electrotaxis device with four electrodes is used for generating non-homogenous electric fields. This device is simple to fabricate and cost effective. It consists of a bottom plate and a top plate. The bottom plate has PDMS channels formed on glass cover slip by soft-lithography technique. NUNC 4-well culture chamber with 4 holes in the center is used as the top plate. Using a thin layer of PDMS the bottom plate and top plate are sealed.
- ImageJ software is used to manually track the cell migrations and the tracks are plotted on a polar plot.
- The dependence of electric field strength on channel dimensions were studied. The *EFSS* was independent of the length and was inversely proportional to width and height of the channel.
- The high resistance in channel heights of more than 1 mm generated Joule heating which increased the temperature inside the chamber by 3 ° C.

- A 52-pin array was inserted into the device chamber to directly measure the *EFSS* in non-homogenous *EFSS* configuration. Configuration 1 had positive and negative electrodes alternated. Configuration 2 had positive and negative electrodes consecutive.
- The simulation and experimental values for homogenous and non-homogenous *EFSS* configurations matched closely with error % of 4 and 2, respectively.
- Simulation data was used to quantify the cell migration speed and direction in the non-homogenous *EFSS* configuration.
- The cells showed good attachment, spreading and elongated distribution of actin filaments with good focal adhesion formations. This proved the biocompatibility of the device.
- Cells showed random migration behavior when no *EFSS* was applied. They migrated at 10 μ/ hr. No change in shape of the cell was observed.
- The cells migrated towards the anode when an *EFSS* of 0.5 V/mm was applied. The cells first elongated and migrated perpendicular to the electric field line direction. The migration speed was noted as 25 μ/ hr.
- In non-homogenous *EFSS* configuration, the cells migrated in the resultant direction of the electric field strength. The cells were not able to distinguish small variation in *EFSS*. The cells migrated at 16 μ/ hr in this configuration.
- By adjusting the *EFSS* in an electrotaxis chamber, the direction of cell migration could be controlled.

7.2 Future work

Studying calcium distribution in the cells in non-homogenous *EFSS* may help us understand the involvement of calcium ions in electrotaxis phenomenon. The cells may be fluorescently labeled for studying calcium distribution in non-homogenous *EFSS*. Previous studies have shown the use of *EFSS* to differentiate stem cells into specific cell types. This device may be used to understand this differentiation behavior more closely. The *EFSS* is well characterized in the device described in this work. The precise control of *EFSS* at every point inside the chamber makes it useful in studying the differentiation behavior of stem cells.

The substrate on which the cells are grown plays an important role in cell behavior. By modifying the surface, the intracellular interactions between nerve cells can be investigated under the influence of *EFSs* for example the axon growth.

References

[1] L. Galvani and G. Aldini. *Aloysii Galvani: De viribus electricitatis in motu musculari.* apud Societatem typographicum, 1792.

[2] E. DuBois-Reymond Untersuchungen ubertierische Elektrizitat II. *Berlin: Reimer,* 1860.

[3] E. Neher and B. Sakmann Single-channel currents recorded from membrane of denervated frog muscle fibres. *Nature* 260: 799 – 802, 1976.

[4] M. Zhao et al. Electrical signals control wound healing through phosphatidylinositol-3-OH kinase-gamma and PTEN. *Nature* 442: 457–460, 2006.

[5] C. McCaig, B. Song and A. Rajnicek Electrical dimensions in cell science. *J Cell Sci* 122 (23): 4267-4276, 2009.

[6] R. Nuccitelli A role for endogenous electric fields in wound healing. *Curr. Topics Develop. Biol.* 58: 1–26, 2003.

[7] R.H.W. Funk and T.K. Monsees Effects of electromagnetic fields on cells: physiological and therapeutical approaches and molecular mechanisms of interaction: A review. *Cells Tissues Organs* 182 (2): 59-78, 2006.

[8] E. Serena, E. Figallo, N. Tandon, C. Cannizzaro, S. Gerecht, N. Elvassore, G. Vunjak-Novakovic Electrical stimulation of human embryonic stem cells: Cardiac differentiation and the generation of reactive oxygen species. *Experim. Cell Research,* 315: 3611–3619, 2009.

[9] L. F. Jaffe and J. W. Vanable Jr. Electric fields and wound healing. *Clin Dermatol.* 2 (3): 34-44, 1984.

[10] B. Song et al. Application of direct current electric fields to cells and tissues in vitro and modulation of wound electric field in vivo. *Nature Protoc.* 2: 1479-1489, 2007.

[11] www.ibidi.com

[12] J. Reece, B. Jane, et al. Campbell Biology. 9th edition; International edition. Harlow: Pearson Education, 2011.

[13] J. A. Madri and D. Graesser Cell migration in the immune system: the evolving inter-related roles of adhesion molecules and proteinases. *Dev. Immunol.* 7 (2-4): 103-16, 2000.

[14] M. Levin Bioelectric mechanisms in regeneration: Unique aspects and future perspectives. Semin. Cell Dev. Biol, 20 (5): 543-56, 2009.

[15] P. Friedl and K. Wolf Tumour-cell invasion and migration: diversity and escape mechanisms. Nat. Rev. Cancer, 3 (5): 362-74, 2003.

[16] A. Lambrechts, M. Van Troys, et al. The actin cytoskeleton in normal and pathological cell motility. *Int J Biochem Cell Biol,* 36 (10): 1890-909, 2004.

[17] M. Vicente-Manzanares, D. J. Webb and A. R. Horwitz Cell migration at a glance. *J Cell Sci,* 118 (21): 4917-4919, 2005.

[18] D. J. Sieg, C. R. Hauck et al. FAK integrates growth-factor and integrin signals to promote cell migration. *Nat Cell Biol,* 2 (5): 249-56, 2000.

[19] N. O. Carragher and M. C. Frame Calpain: a role in cell transformation and migration. *Int J Biochem Cell Biol,* 34 (12): 1539-43, 2002.

[20] W. A. Catterall Voltage-gated calcium channels. *Cold Spring Harb Perspect Biol.,* 3 (8), 2011.

[21] A. Schwab, V. Nechyporuk-Zloy et al. Cells move when ions and water flow. *Pflugers Arch,* 453 (4): 421-32, 2007.

[22] E. K. Onuma and S.-W. Hui Electric field-directed cell shape changes, displacement, and cytoskeletal reorganization are calcium dependent. *J. Cell Biol.* 106: 2067-2075, 1988.

[23] M. S. Cooper and R. E. Keller Perpendicular orientation and directional migration of amphibian neural cells in dc electrical fields. *Proc. Natl. Acad. Sci. USA* 81: 160-164, 1984.

[24] N. Ozkucur, T. K. Monsees, S. Perike, H. Q. Do, R. H.W. Funk Local Calcium Elevation and Cell Elongation Initiate Guided Motility in Electrically Stimulated Osteoblast-Like Cells. *PLoS One,* 4 (7): 6131-6139, 2009.

[25] R. Nuccitelli, T. Smart and J. Ferguson Protein kinases are required for embryonic neural crest cell galvanotaxis. *Cell Motil. Cytoskeleton,* 24: 54-66, 1993.

[26] H. Ishibashi, A. Dinudom, K. F. Harvey, S. Kumar, J. A. Young and D. I. Cook Na(+)-H(+) exchange in salivary secretory cells is controlled by an intracellular Na(+) receptor. *Proc. Natl. Acad. Sci. USA* 96: 9949-9953, 1999.

[27] M. Zhao, J. Pu, J. V. Forrester and C. D. McCaig Membrane lipids, EGF receptors, and intracellular signals colocalize and are polarized in epithelial cells moving directionally in a physiological electric field. *FASEB J.* 16: 857-859, 2002.

[28] R. Nuccitelli Endogenous electric fields in embryos during development, regeneration and wound healing. *Rad Protect Dosimetry,* 106 (4): 375–383, 2003.

[29] M. Verworn Unter- suehungen fiber die polare Erregung der lebendigen Substanz durch den konstanten Strom III. *Pflugers Arch,* 62: 415-450, 1896.

[30] O. Carlgren ber die Einwirkung des konstanten galvanischcn Stromes auf niedere Organismen. *Arch. f. Physiol,* 49-75, 1900.

[31] L. Hirsehfeld Ein Versuch einige Lebeuserseheinungender Amoeben phyikalisehehemisch zu erklaeren. *Z allg Physiol,* 9: 529—534, 1909.

[32] J. J. McClendon Ein Versuch am Sboide Bewegung als Folgeerscheinung wechselnden elektrischen Polarisationszustandes der Plasmahaut zu erklaeren. *Pflueigers Arch*, 140: 271-280, 1911.

[33] R. H. Lace Orientation to the electric current and to light in *Amoeba. Anat Ree*, 32: 55, 1926

[34] A. W. Greeley Experiments on the physical structure of the protoplasm of *Paramecium* and its relation to the reactions of the organism to thermal, chemical and electrical stimuli. *Biol Bull Mar biol Labor*, 7: 3-32, 1904.

[35] S. O. Mast Structure, movement, locomotion and stimulation in *Amoeba. J Morph a Physiol*, 41: 347-425, 1926.

[36] J. A. Miller and L. S. Goldston Galvanotropic responses of paramecia to balanced square waves, *Ohio Journal of Science*, 47(3): 127-129, 1947.

[37] A. Watanabe, M. Kodati and S. Kinoshita Uber die negative Galvanotaxis der Myxomyceten-Plasmodien, *Bot. Mag.*, Tokyo, 52: 441-449, 1938.

[38] J. D. Anderson Galvanotaxis of slime molds. J. general Physio, 1-16, 1951.

[39] Fukushima et al [39] in 1953 showed that the granulocytes move towards the anode at high pH values.

[40] L. F. Jaffe and R. Nuccitelli An ultra sensitive vibrating probe for measuring steady extracellular currents *J. Cell Biol.*, 63: 614-28, 1974.

[41] R. B. Borgens, J. W. Vanable Jr. and L. F. Jaffe Bioelectricity and regeneration: Large currents leave the stumps of regenerating newt limbs. *Proc. Nat. Acad. Sci.*, 74: 4528-4532, 1977.

[42] R. B. Borgens, J. W. Vanable Jr. and L. F. Jaffe Bioelectricity and regeneration: I. Initiation of frog limb regeneration by minute currents. *J. Exp. Zool.* 200: 403-16, 1977.

[43] C. M. Illingworth and A. T. Barker Measurement of electrical currents during regeneration of amputated finger tips in children, *Clin. Phys. Physiol. Meas.* 1 (87): , 1980.

[44] C. A. Erickson and R. Nuccitelli, Embryonic fibroblast motility and orientation can be influenced by physiological electric fields, *J. Cell Biol.*, 98: 296–307, 1984.

[45] B. Rapp, A. de Boisfleury-Chevance, H. Gruler Galvanotaxis of human granulocytes. Dose-response curve, *Eur Biophys J.*, 16 (5): 313-319, 1988.

[46] A. M. Rajnicek, N. A. Gow and C. D. McCaig Electric field-induced orientation of rat hippocampal neurones in vitro. *Exp Physiol.*, 77 (1): 229-32, 1992.

[47] R. Nuccitelli and T. Smart Extracellular calcium levels strongly influence neural crest cell galvanotaxis. *Biol. Bull*, 176: 130–135, 1989.

[48] A. T. Barker, L.F. Jaffe and J. W. Vanable Jr. The glabrous epidermis of cavies contains a powerful battery. *Am J Physiol*, 242: 348-366, 1982.

[49] K. Y. Nishimura, R. R. Isseroff, R. Nuccitelli Human keratinocytes migrate to the negative pole in direct current electric fields comparable to those measured in mammalian wounds. *J Cell Sci.*, 109 (1): 199-207, 1996.

[50] D. M. Sheridan, R. R. Isseroff and R. Nuccitelli Imposition of a physiologic DC electric field alters the migratory response of human keratinocytes on extracellular matrix molecules. *J Invest Dermatol.*, 106 (4): 642-6, 1996.

[51] H. Gruler and R. Nuccutelli Neural crest cell galvanotaxis: new data and a novel approach to the analysis of both galvanotaxis and chemotaxis. *Cell Motil Cytoskeleton,* 19(2): 121-33, 1991.

[52] R. Karba, D. Šemrov, L. Vodovnik, H. Benko and R. Savrin DC electrical stimulation for chronic wound healing enhancement Part 1. Clinical study and determination of electrical field distribution in the numerical wound model. *Bioelectrochemistry and Bioenergetics*, 43 (2): 265-270, 1997.

[53] P. Martin Wound healing--aiming for perfect skin regeneration. *Science*, 276 (5309): 75-81, 1997.

[54] S. E. Gardner, R. A. Frantz and F. L. Schmidt Effect of electrical stimulation on chronic wound healing: a meta-analysis. *Wound Repair Regen.* 7 (6): 495-503, 1999.

[55] M. Zhao, A. Agius-Fernandez, J. V. Forrester and C. D. McCaig Directed migration of corneal epithelial sheets in physiological electric fields. *Invest Ophthalmol Vis Sci.* 37 (13): 2548-58, 1996.

[56] M. Zhao, A. Agius-Fernandez, J. V. Forrester and C. D. McCaig Orientation and directed migration of cultured corneal epithelial cells in small electric fields are serum dependent. *J Cell Sci.,* 109 (6): 1405-14, 1996.

[57] M. Zhao, C.D. McCaig, A. Agius-Fernandez, J. V. Forrester and K. Araki-Sasaki Human corneal epithelial cells reorient and migrate cathodally in a small applied electric field. *Curr Eye Res.* 16 (10): 973-984, 1997

[58] B. Farboud, R. Nuccitelli, I. R. Schwab and R. R. Isseroff DC Electric Fields Induce Rapid Directional Migration in Cultured Human Corneal Epithelial Cells. *Exp. Eye Res.* 70:667-673, 2000.

[59] W. Korohoda, M. Mycielska, E. Janda and Z. Madeja Immediate and Long-Term Galvanotactic Responses of Amoeba proteus to dc Electric Fields. *Cell Motility and the Cytoskeleton,* 45:10–26, 2000.

[60] Z. Sayers, A. M. Roberts and L. H. Bannister Random walk analysis of movement and galvanotaxis of Amoeba proteus. *Acta Proto zool*, 18: 313-325, 1979.

[61] C. E. Pullar, R. R. Isseroff and R. Nuccitelli Cyclic AMP-Dependent Protein Kinase A Plays a Role in the Directed Migration of Human Keratinocytes in a DC Electric Field. *Cell Motility and the Cytoskeleton*, 50:207–217, 2001.

[62] M. B. A. Djamgoz, M. Mycielska, Z. Madeja, S. P. Fraser and W. Korohoda Directional movement of rat prostate cancer cells in direct-current electric field: involvement of voltage-gated Na+ channel activity. *J. Cell Sci*, 114 (14): 2697-2705, 2001.

[63] K. S. Fang, B. Farboud, R. Nuccitelli and R. R. Isseroff Migration of human keratinocytes in electric fields requires growth factors and extracellular calcium. *J Invest Dermatol.* 111 (5): 751-756, 1998.

[64] D. R. Trollinger, R. R. Isseroff and R. Nuccitelli Calcium channel blockers inhibit galvanotaxis in human keratinocytes. *J Cell Physio.*, 193: 1–9, 2002.

[65] E. Wang, M. Zhao, J. V. Forrester and C. D. McCaig Electric fields and map kinase signaling can regulate early wound healing in lens epithelium. *IOVS,* 44 (1): 244-249, 2003.

[66] M. E. Mycielska and M. B. A. Djamgoz Cellular mechanisms of direct-current electric field effects: galvanotaxis and metastatic disease. *J Cell Sci*, 117: 1631-1639, 2004.

[67] S. Sun, J. Wise, and M. Cho Human fibroblast migration in three-dimensional collagen gel in response to noninvasive electrical stimulus. *Tissue Engg*, 10 (9/10): 1548-1557, 2004.

[68] B. Reid, B. Song, C. D. McCaig and M. Zhao Wound healing in rat cornea: the role of electric currents. *FASEB J.*, 19(3): 379–386, 2005.

[69] N. Ogawa, H Okua, K. Hashimoto, M. Ishikawa A physical model for galvanotaxis of Paramecium cell. *Journal of Theoretical Biology*, 242: 314–328, 2006.

[70] D. M. Brunette Spreading and orientation of epithelial cells on grooved substrata. *Exp Cell Res,* 167: 203–217, 1986.

[71] Pen-h. G. Chao, H. H. Lu, C. T. Hung, S. B. Nicoll and J. C. Bulinski Effects of applied dc electric field on ligament fibroblast migration and wound healing. *Connective Tissue Research*, 48: 188–197, 2007

[72] E. I. Finkelstein, Pen-h. G. Chao, C. T. Hung and J. C. Bulinski Electric Field-Induced Polarization of Charged Cell Surface Proteins Does Not Determine the Direction of Galvanotaxis. *Cell Mot Cytoskel*, 64: 833–846, 2007.

[73] M. J. Sato, M. Ueda, H. Takagi, T. M. Watanabe, T. Yanagida , M. Ueda Input–output relationship in galvanotactic response of *Dictyostelium* cells. *BioSystems*, 88: 261–272, 2007.

[74] R. Nuccitelli, P. Nuccitelli, S. Ramlatchan, R. Sanger and P. J. S. Smith Imaging the electric field associated with mouse and human skin wounds. *Wound Rep Reg*, 16: 432–441, 2008.

[75] F. Lin, F. Baldessari, C. C. Gyenge, T. Sato, R. D. Chambers, J. G. Santiago and E. C. Butcher Lymphocyte electrotaxis in vitro and in vivo. *J Immun*, 2465-2471, 2008.

[76] www.invitrogen.com

[77] G. Yang, H. Long, J. Wu, H. Huang A novel electrical field bioreactor for wound healing study. *2008 International Conference on BioMedical Engineering and Informatics.*

[78] C-W. Huang, J-Y. Cheng, M-H. Yen, T-H. Young Electrotaxis of lung cancer cells in a multiple-electric-field chip. *Biosensors and Bioelectronics*, 24: 3510–3516, 2009.

[79] C-C. Wang, Y-C. Kao, P-Y. Chi, C-W. Huang, J-Y. Lin, C-F. Chou, J-Y. Cheng and C-H. Lee Asymmetric cancer-cell filopodium growth induced by electric-fields in a microfluidic culture chip. *Lab Chip*, 11: 695–699, 2011.

[80] P. Rezai, A. Siddiqui, P. R. Selvaganapathy and B. P. Gupta Electrotaxis of Caenorhabditis elegans in a microfluidic environment. Lab Chip, 10: 220–226, 2010.

[81] J. Li, S. Nandagopal, D. Wu, S. F. Romanuik, K. Paul, D. J. Thomson and F. Lin Activated T lymphocytes migrate toward the cathode of DC electric fields in microfluidic devices. *Lab Chip*, 11:1298-1304, 2011.

[82] X. Maniere, F. Lebois, I. Matic, B. Ladoux, J-M. D. Meglio, P. Hersen Running Worms: C. elegans Self-Sorting by Electrotaxis. *PLoS ONE*, 6(2): 1-7, 2011.

[83] Z. Zhao, C. Watt, A. Karystinou, A. J. Roelofs, C. D. McCaig, I. R. Gibson and C. D. Bari Directed migration of human bone marrow mesenchymal stem cells in a physiological direct current electric field. *European Cells and Materials*, 22: 344-358, 2011.

[84] J-F. Feng, J. Liu, X-Z. Zhang, L. Zhang, J-Y. Jiang, J. Nolta, M. Zhao Brief report: Guided migration of neural stem cells derived from human embryonic stem cells by an electric field. *Stem cells*, 30: 349–355, 2012.

[85] D. C. Duffy, J. C. McDonald, O. J. A. Schueller and G. M. Whitesides. Rapid prototyping of microfluidic systems in polydimethylsiloxane. *Anal Chem*, 70: 4974-4984, 1998.

[86] J. C. McDonald, D. C. Duffy, J. R. Anderson, D. T. Chiu, H. Wu, O. J. A. Schueller and G. M. Whitesides. Fabrication of microfluidic systems in polydimethylsiloxane. *Electrophoresis*, 21: 27-40, 2000.

[87] J. M. K. Ng, G. I. Gitlin, A. D. Stroock and G. M. Whitesides. Components for integrated poly(dimethylsiloxane microfluidic systems. *Electrophoresis*, 23: 3461-3473, 2002.

[88] G. J. M. Fechine, M. S. Rabello, R. M. Souto Maior and L. H. Catalani. Surface characterization of photodegraded poly (ethylene terephthalate): The effect of ultraviolet absorbers. *Polymer*, 45(7): 2303–2308, 2004.

[89] M. Berek. Grundlagen der Tiefenwahrnehmung im Mikroskop. *Marburg Sitzungsberichte*, 62: 189–223, 1927.

[90] J. Li and F. Lin. Microfluidic devices for studying chemotaxis and electrotaxis. *Trends in Cell Biology*, 21 (8): 489-497, 2011.

[91] F. Gielen, F. Pereira, A. J. Demello. High-resolution local imaging of temperature in dielectrophoretic platforms. *Anal Chem,* 82: 7509-7514, 2010.